Sustainable Protein Sources: Extraction, Structural Characterization, and Functional Properties

Sustainable Protein Sources: Extraction, Structural Characterization, and Functional Properties

Guest Editors

Armin Mirzapour-Kouhdasht
Jen-Yi Huang

MDPI

Basel • Beijing • Wuhan • Barcelona • Belgrade • Novi Sad • Cluj • Manchester

Guest Editors
Armin Mirzapour-Kouhdasht
Department of Food Science
Purdue University
West Lafayette, IN
USA

Jen-Yi Huang
Department of Food Science
Purdue University
West Lafayette, IN
USA

Editorial Office
MDPI AG
Grosspeteranlage 5
4052 Basel, Switzerland

This is a reprint of the Special Issue, published open access by the journal *Foods* (ISSN 2304-8158), freely accessible at: https://www.mdpi.com/journal/foods/special_issues/A4WMEVA0F2.

For citation purposes, cite each article independently as indicated on the article page online and as indicated below:

Lastname, A.A.; Lastname, B.B. Article Title. *Journal Name* **Year**, *Volume Number*, Page Range.

ISBN 978-3-7258-7759-1 (Hbk)
ISBN 978-3-7258-7760-7 (PDF)
https://doi.org/10.3390/books978-3-7258-7760-7

Contents

About the Editors

Armin Mirzapour-Kouhdasht

Armin Mirzapour-Kouhdasht is a Postdoctoral Research Associate in the Department of Food Science and Technology at Ohio State University, United States. His research interests center on sustainable protein sources, bioactive peptides, food processing technologies, and the structure function relationships of proteins derived from plant, marine, algal, and microbial origins. He has held postdoctoral and visiting research positions at leading institutions in the United States, Ireland, the United Kingdom, and South Korea, contributing to interdisciplinary projects on protein extraction, functional food development, waste valorization, and advanced processing technologies. He has authored numerous peer reviewed publications and book chapters in high-impact journals and has received competitive research funding for projects focused on sustainable biomaterials and protein innovation.

Jen-Yi Huang

Jen-Yi Huang is the Dale A. Seiberling Endowed Professor in the Departments of Food Science and Technology and Food, Agricultural and Biological Engineering at Ohio State University. His expertise lies in innovative food processing technologies, including microbubble, cold plasma, electrodialysis, etc., for improving food quality and safety with reduced energy and water use. He is also interested in transformation of our food systems through life cycle approaches to enhance their environmental sustainability. His research has been funded by federal agencies and industries for totaling nearly $13 million USD, and has led to publication of 61 peer-reviewed papers.

Preface

This Special Issue Reprint examines sustainable protein sources that help meet the growing global need for high quality dietary protein while minimizing environmental issues. The Reprint shows the methods used to extract proteins from sustainable, renewable sources, which include plant materials, microalgae, fungi, and microbial biomass. Researchers investigate how structural elements of proteins interact with their functional properties which develop through different processing techniques and modification methods.

The Reprint unites recent scientific discoveries that enable better protein functionality through improved digestibility, bioavailability, and the safety of protein products for food applications. The collected works highlight emerging extraction methods, novel hybrid and conjugated protein systems, and applications in areas such as food formulation, meat analogs, and functional foods. The Reprint presents research in protein science that applies to usable research to develop sustainable protein ingredients that provide nutritional benefits.

The Reprint exists because researchers need to acquire diverse protein sources, which can help them develop new food products in scientific research. The included studies show current research directions while presenting results that help both universities and businesses create future developments.

This Reprint is addressed to researchers, food scientists, technologists, and industry professionals working in protein chemistry, food processing, product development, and sustainable food systems.

Armin Mirzapour-Kouhdasht and Jen-Yi Huang
Guest Editors

foods

Editorial

Sustainable Protein Sources: Extraction, Structural Characterization, and Functional Properties

Armin Mirzapour-Kouhdasht [1,*] and Jen-Yi Huang [1,2]

[1] Department of Food Science and Technology, The Ohio State University, Columbus, OH 43210, USA; huang.6338@osu.edu
[2] Department of Food, Agricultural, and Biological Engineering, The Ohio State University, Columbus, OH 43210, USA
* Correspondence: mirzapourkouhdasht.1@osu.edu

The worldwide demand for dietary protein has risen, while environmental problems and sustainability issues with traditional animal protein production methods have increased, accelerating research efforts to explore sustainable protein sources and alternative protein sources. The Special Issue "Sustainable Protein Sources: Extraction, Structural Characterization and Functional Properties" presents recent research developments that provide solutions to essential protein extraction methods, protein structural understanding, and food system functional performance issues (Contributions 1–8).

Sustainable protein extraction methods face a prominent developmental hurdle to overcome before they can be put into use by researchers, who are developing extraction methods to provide the highest possible yields of protein while still preserving nutritional and functional values. The Special Issue provides several papers describing novel processes to extract and process materials using some of the following methods: enzymatic-assisted extraction, mechanical disruption techniques, extrusion-based processing, and optimized alkaline solubilization. The results presented reveal that customized extraction techniques will produce optimal results, while reducing the amount of energy required for production, therefore allowing for the development of more environmentally friendly and more viable protein manufacturing processes (Contributions 1–4).

In addition to extraction, it is also critical to understand how proteins are structured so that they will perform satisfactorily once added to a food product. The studies in this Special Issue utilize advanced techniques to further characterize physicochemical properties to understand how processing, fractionating, and/or modifying proteins will alter their conformational properties and intermolecular interactions. The findings from these studies collectively indicate that changes to protein secondary and tertiary structures fundamentally impact the protein's performance characteristics, such as solubility, emulsification capacity, foam stability, and gelation (Contributions 3, 6 and 7).

In addition to techno-functional performance, nutritional quality and biological effects must also be actively considered for sustainable protein adoption. Several articles published in this Special Issue examine digestibility, bioavailability, and bioactivity of sustainable proteins and evaluate potential safety concerns (e.g., allergenicity, cytotoxicity). Strategies (e.g., germination, enzymatic hydrolysis, conjugation, protein blending) that enhance digestibility/nutritional quality, while reducing undesirable biological effects, have been identified in these studies, reinforcing the need for evaluating nutrition/safety during the protein creation pipeline (Contributions 2, 5 and 7). Various studies have produced comparable outcomes on different alternative protein sources, highlighting the impact of the processing method used to produce these proteins on their digestibility and health benefits.

 https://doi.org/10.3390/foods15081357

Such examples of protein processing methods are ultrasound-assisted extraction, enzyme hydrolysis, and protein–phenol conjugation. Evidence supports the improved digestibility of and reduced antinutritional factors in protein from plants as a consequence of using these processing methods [1–4]. Studies performed using insect-based protein sources and pseudocereal protein sources demonstrate similar results of increased nutritional quality and functional characteristics as a result of using germination and protein blending methods [5–8]. Collectively, these findings on alternative proteins identify the necessity of conducting an integrated evaluation of the proteins' nutritional and safety aspects in conjunction with the techno-functional characteristics of the completed product throughout the entire production process of sustainable sources of protein.

The different sources of protein from multiple areas that are investigated within the studies printed in this Special Issue demonstrate the wide-ranging scope of the present sustainability sciences of protein research program. The studies consist of plant proteins, microalgal proteins, fungus-derived proteins, and hybrid systems utilizing proteins from various sources. Many of the studies review the application of these proteins in complex food matrices, including meat alternatives or analogues, protein blends, and structured food systems, highlighting the future potential of using these sustainable protein sources as a replacement for or complement to traditional protein sources in future food products (Contributions 3, 4 and 8).

There are still significant gaps in knowledge; however, this Special Issue documents substantial progress in research on sustainable proteins. Future research will need to focus on developing processing technologies that are environmentally sustainable, scalable, and low-cost, in addition to creating validated methodologies for characterizing proteins and evaluating protein functionality. In addition, there needs to be more focus on linking macroscopic and microscopic food properties to consumer attributes. Furthermore, integrating sustainable proteins into new food technology, such as advanced formulation methods and structured food design, presents an excellent future research opportunity (Contributions 6–8).

Overall, this Special Issue has provided a thorough review of new advances in sustainable protein research, which includes extracting technologies, the structural characterization of proteins, their functional performance, and the nutritional aspects. Therefore, the consolidation of the advances presented in this Special Issue will create a more thorough understanding of how to best develop and use sustainable proteins within food systems. We hope this Special Issue will serve as an important resource and inspiration for continued cross-disciplinary research focused on the development of sustainable protein solutions that will help build a resilient and secure global food supply.

Author Contributions: Conceptualization, A.M.-K. and J.-Y.H.; writing—original draft preparation, A.M.-K. and J.-Y.H.; review and editing A.M.-K. and J.-Y.H. All authors have read and agreed to the published version of the manuscript.

Funding: This research received no external funding.

Data Availability Statement: Not applicable.

Conflicts of Interest: The authors declare no conflicts of interest.

List of Contributions:

1. Ghosh, S.; González Hernández, E.C.; Sha, X.; Chow, J.; San Martin-Gonzalez, F.; Jin, Q.; Chen, D. Rapid Protein Extraction from Canola Meal Pre-Treated with Enzymatic Reactive Extrusion. *Foods* **2026**, *15*, 498. https://doi.org/10.3390/foods15030498.

2. Szołtysik, M.; Mandecka, A.; Maciejewska, M.; Dąbrowska, A.; Nowak, M. Functional, Biological and Nutritional Properties of Protein Fraction Isolated from *Yarrowia lipolytica* Biomass. *Foods* **2025**, *14*, 3801. https://doi.org/10.3390/foods14213801.
3. Khalesi, M.; Dowling, S.; Comerford, J.; Sweeney, C.; Esteghlal, S.; FitzGerald, R.J. Emulsification Properties of Plant and Milk Protein Concentrate Blends. *Foods* **2025**, *14*, 3406. https://doi.org/10.3390/foods14193406.
4. Ng, J.W.; Lee, S.Y.; Teh, T.M.; Weingarten, M.; Talukder, M.M.R. Enhanced Protein Extraction from *Auxenochlorella protothecoides* Through Synergistic Mechanical Cell Disruption and Alkaline Solubilization. *Foods* **2025**, *14*, 2597. https://doi.org/10.3390/foods14152597.
5. Li, N.; Li, H.; Feng, D.; Li, M.; Han, D.; Liu, T.; Wang, J. Anti-Diabetic Effect of Soy–Whey Dual-Protein on Mice with Type 2 Diabetes Mellitus Through INS/IRS1/PI3K Signaling Pathway. *Foods* **2025**, *14*, 2115. https://doi.org/10.3390/foods14122115.
6. Hashemi, H.; Eskandari, M.H.; Khalesi, M.; Golmakani, M.-T.; Niakousari, M.; Hosseini, S.M.H. Effects of Conjugation with Basil Seed Gum on Physicochemical, Functional, Foaming, and Emulsifying Properties of Albumin, Whey Protein Isolate and Soy Protein Isolate. *Foods* **2025**, *14*, 390. https://doi.org/10.3390/foods14030390.
7. Mirzapour-Kouhdasht, A.; Shaghaghian, S.; Majdinasab, M.; Huang, J.-Y.; Garcia-Vaquero, M. Unravelling the Digestibility and Structure–Function Relationship of Lentil Protein Through Germination and Molecular Weight Fractionation. *Foods* **2025**, *14*, 272. https://doi.org/10.3390/foods14020272.
8. Santhapur, R.; Jayakumar, D.; McClements, D.J. Formation and Characterization of Mycelium–Potato Protein Hybrid Materials for Application in Meat Analogs or Substitutes. *Foods* **2024**, *13*, 4109. https://doi.org/10.3390/foods13244109.

References

1. Aghababaei, F.; McClements, D.J.; Hadidi, M. Ultrasound processing for enhanced digestibility of plant proteins. *Food Hydrocoll.* **2024**, *155*, 110188. [CrossRef]
2. Gu, J.; Bk, A.; Wu, H.; Lu, P.; Nawaz, M.A.; Barrow, C.J.; Suleria, H.A.R. Impact of processing and storage on protein digestibility and bioavailability of legumes. *Food Rev. Int.* **2023**, *39*, 4697–4724. [CrossRef]
3. Yan, X.; Zeng, Z.; McClements, D.J.; Gong, X.; Yu, P.; Xia, J.; Gong, D. A review of the structure, function, and application of plant-based protein–phenolic conjugates and complexes. *Compr. Rev. Food Sci. Food Saf.* **2023**, *22*, 1312–1336. [CrossRef] [PubMed]
4. Liu, J.; Yong, H.; Yao, X.; Hu, H.; Yun, D.; Xiao, L. Recent advances in phenolic–protein conjugates: Synthesis, characterization, biological activities and potential applications. *RSC Adv.* **2019**, *9*, 35825–35840. [CrossRef] [PubMed]
5. Jin, J.; Ohanenye, I.C.; Udenigwe, C.C. Buckwheat proteins: Functionality, safety, bioactivity, and prospects as alternative plant-based proteins in the food industry. *Crit. Rev. Food Sci. Nutr.* **2022**, *62*, 1752–1764. [CrossRef] [PubMed]
6. Karaca, A.C.; Nickerson, M.; Caggia, C.; Randazzo, C.L.; Balange, A.K.; Carrillo, C.; Capanoglu, E. Nutritional and functional properties of novel protein sources. *Food Rev. Int.* **2023**, *39*, 6045–6077. [CrossRef]
7. Queiroz, L.S.; Silva, N.F.N.; Jessen, F.; Mohammadifar, M.A.; Stephani, R.; de Carvalho, A.F.; Casanova, F. Edible insect as an alternative protein source: A review on the chemistry and functionalities of proteins under different processing methods. *Heliyon* **2023**, *9*, e14831. [CrossRef] [PubMed]
8. Huang, C.; Feng, W.; Xiong, J.; Wang, T.; Wang, W.; Wang, C.; Yang, F. Impact of drying method on the nutritional value of the edible insect protein from black soldier fly (*Hermetia illucens* L.) larvae: Amino acid composition, nutritional value evaluation, in vitro digestibility, and thermal properties. *Eur. Food Res. Technol.* **2019**, *245*, 11–21. [CrossRef]

foods

MDPI

Article

Formation and Characterization of Mycelium–Potato Protein Hybrid Materials for Application in Meat Analogs or Substitutes

Ramdattu Santhapur, Disha Jayakumar and David Julian McClements *

Department of Food Science, University of Massachusetts, Amherst, MA 01003, USA; rsanthapur@umass.edu (R.S.); djayakumar@umass.edu (D.J.)
* Correspondence: mcclements@foodsci.umass.edu; Tel.: +1-413-545-2275

Abstract: There is increasing interest in the development of meat analogs due to growing concerns about the environmental, ethical, and health impacts of livestock production and consumption. Among non-meat protein sources, mycoproteins derived from fungal fermentation are emerging as promising meat alternatives because of their natural fibrous structure, high nutritional content, and low environmental impact. However, their poor gelling properties limit their application in creating meat analogs. This study investigated the potential of creating meat analogs by combining mycoprotein (MCP), a mycelium-based protein, with potato protein (PP), a plant-based protein, to create hybrid products with meat-like structures and textures. The PP-MCP composites were evaluated for their physicochemical, rheological, textural, and microstructural properties using electrophoresis, differential scanning calorimetry, dynamic shear rheology, texture profile analysis, confocal fluorescence microscopy, and scanning electron microscopy analyses. The PP-MCP hybrid gels were stronger and had more fibrous structures than simple PP gels, which was mainly attributed to the presence of hyphae fibers in mycelia. Dynamic shear rheology showed that the PP-MCP hybrids formed irreversible heat-set gels with a setting temperature of around 70 °C during heating, which was attributed to the unfolding and aggregation of the potato proteins. Confocal and electron microscopy analyses showed that the hybrid gels contained a network of mycelia fibers embedded within a potato protein matrix. The hardness of the PP-MCP composites could be increased by raising the potato protein content. These findings suggest that PP-MCP composites may be useful for the development of meat analogs with more meat-like structures and textures.

Keywords: potato protein; mycelium; biopolymer composites; alternative proteins; hybrid products

1. Introduction

Alternative protein sources are increasingly being used to formulate foods to mitigate the adverse environmental impacts associated with the production of animal proteins, especially meat [1]. At around 14–18% of the total, livestock production is a major contributor to global greenhouse gas emissions (GHGs). Deforestation for grazing pasture, fertilizer use for feed crops, and methane from ruminant digestion have been reported to be the main causes of GHGs in this sector [2–4]. Animal husbandry also utilizes considerable amounts of land and water, which exacerbates biodiversity loss and environmental degradation. Switching to alternative protein sources could greatly reduce these impacts, slowing climate change and advancing global sustainability [5,6]. Researchers have reported that substituting conventional meat products for plant-based, mycoprotein, or cultured meat products could reduce agricultural GHG emissions by over 50% by 2050 in high-consumption areas [7,8]. Alternative protein sources also have the potential to improve public health by reducing the risks of diet-related diseases that have been linked to excessive meat intake, such as cancer and cardiovascular disease [9,10]. Thus, alternative proteins have considerable potential for reaching global climate goals and promoting human health. Nevertheless, there are advantages and disadvantages associated with each individual kind of alternative protein source for formulating meat analogs and substitutes. For example, plant proteins are relatively abundant and affordable, but it is often challenging to accurately mimic the fibrous structure of real meat products using them, which reduces consumer acceptance [11,12]. In contrast, mycoproteins can better mimic the fibrous structure of meat products, but they are more difficult to produce affordably on a large scale and they cannot form strong gels when used alone [13]. Consequently, it is beneficial to create hybrid products that combine the benefits of these two alternative protein sources.

Mycoproteins are becoming popular for formulating meat analogs or substitutes because of their high protein content, positive health benefits, and low environmental impact [14]. This source of alternative proteins is usually produced in bioreactors by fungal fermentation, and then isolated and purified before being converted into meat substitutes [15,16]. Compared to animal proteins, mycoprotein has a far smaller carbon footprint and uses less energy and water. Indeed, the carbon footprint of mycoprotein is around ten times lower than that of beef and four times lower than that of chicken. A recent environmental and nutritional life cycle analysis of different protein-rich food sources found that global warming potential, water use, land use, and pollution were much less for mycoprotein than for beef, appreciably less than for chicken, and appreciably more than for plant proteins (tofu) [17]. The higher impact of mycoprotein-based products on the environment than plant protein-based ones has been attributed to the relatively high amounts of energy and raw materials required to run the bioreactors used to generate the mycelia [17]. Mycoprotein-based food products are rich in proteins, dietary fibers, vitamins, and minerals, and low in fat, which may lead to nutritional and health benefits [15]. Indeed, consumption of mycoproteins has been reported to stimulate the insulin response, promote digestive health, increase satiety, reduce cholesterol levels, and modulate blood sugar levels [15,16,18]. Consequently, mycelium-based food products have great potential to contribute to the world's growing protein requirements [18].

Plant proteins are another important alternative protein source used to formulate meat analogs and substitutes [19–21]. Potato protein is a non-allergenic source of plant proteins [22]. Potatoes typically have a relatively low total protein content (1 to 2%). However, there are large quantities of protein-rich wastewater produced by the potato processing industry that can be converted into a value-added nutritional and functional ingredients for application in foods and other products [23]. The proteins isolated from potatoes are mainly comprised of patatin (35–40%) and protease inhibitors (30–40%) [22]. Potato proteins are considered to be a good nutritional source of proteins because they contain all the essential amino acids required to maintain human health and wellbeing [24,25]. Moreover, potato proteins can form strong heat-set gels because the protein molecules are typically in a native state in commercial ingredients [22,26,27], which is often not the case for more common commercial plant protein ingredients, such as those isolated from soybeans or peas [28,29].

In our previous study, we examined the possibility of creating hybrid products by combining mushrooms with either whey proteins [30] or potato proteins [31]. These studies showed that the introduction of the mushroom extracts into the protein gels created hybrid products that had a more fibrous structure than the original protein gels. However, the presence of the mushroom extracts could either increase or decrease the mechanical strength of the protein gels, depending on the system. Consequently, more research is needed to determine the factors that impact the structural and mechanical properties of hybrid products formed from different sources of alternative proteins. In the current study, we examined the properties of hybrid food products containing mycoproteins and potato proteins. The mycoproteins were selected because of their ability to create fibrous structures and their good nutritional profile, while the potato proteins were selected because of their ability to form firm heat-set gels and their high protein content. In particular, we examined how the appearance, rheology, texture, and microstructure of these hybrid products depended on their composition. It should be noted that other researchers have recently examined the impact of mineral ions on the properties of mycelium–potato protein hybrids formulated from a different kind of mycelium, i.e., *Fusarium venenatum* [32]. The results of this study suggested that electrostatic interactions played a key role in the formation and properties of these hybrids.

The information obtained in our study may be useful for creating more sustainable alternatives to meat products.

2. Materials and Methods

2.1. Materials

Mycelium was generously provided by the Better Meat Co. (Sacramento, CA, USA). Potato protein ("PP200," Solanic®200) was kindly supplied by Royal Avebe (Veendam, The Netherlands). The company specified that the powdered potato protein ingredient contained 90.5% protein, 7.0% water, 2.9% salt, less than 0.2% carbohydrates, and less than 0.1 g fat (*w*/*w*). Sodium chloride (NaCl) was obtained from the Sigma-Aldrich Co. (St. Louis, MO, USA).

2.2. Mycoprotein Preparation

The mycoprotein samples were initially provided by the manufacturer in the form of dried pellets. Proximate analysis of these mycoprotein samples was carried out using standard AOAC methods, as described in our previous study on mushrooms [31]. The mycoprotein pellets were then converted into a fine powder using a food processor operated at 6000 rpm for 90 s. This powder was then sieved through a 20 mm mesh to achieve a uniform particle size. Moreover, sieving removed any large insoluble aggregates that might have sedimented during the preparation of the gels, thereby leading to inhomogeneous samples and inconsistent results. The powder was then hydrated in deionized water for 4 h before being blended into a paste using a meat blender. These samples were then utilized to prepare the hybrid products used for the rheological, textural, and microstructural tests.

2.3. Preparation of Mycoprotein–Potato Protein Hybrids

Potato protein (PP) and mycoprotein (MCP) powders were blended together in various ratios and then dispersed in double-distilled water. The mixtures were then stirred overnight at 350 rpm using a magnetic stirrer to ensure proper hydration and dispersion. A series of test samples was then prepared based on their ability to form heat-set gels: (i) *potato protein samples*: 5%, 10%, and 15% PP; (ii) *mycoprotein samples*: 15% MCP; and (iii) *hybrid samples*: 15% MCP with either 10% or 15% PP. Some dispersions were analyzed directly for zeta potential, differential scanning calorimetry, and dynamic shear rheology, while others were converted into heat-set gels by heating then in a water bath at 90 °C for 30 min and then used for texture profile and microscopy analyses. The PP and MCP concentrations used in this study were selected with the aim of creating final hybrid products with a protein level, texture, and structure somewhat similar to that found in real meat.

2.4. Zeta Potential Analysis

Microelectrophoresis (Zetasizer Nano ZS, Malvern Instruments, Malvern, Worcestershire, UK) was used to characterize the electrical properties of the potato protein, mycoprotein, and hybrid samples. This method is based on measuring the direction and speed of movement of molecules or particles within an oscillating applied electric field [33]. First, dispersions of 0.1% (*w*/*v*) of PP and/or MCP were prepared with pH values ranging from 3 to 8. The zeta-potential values of these dispersions were then measured using the microelectrophoresis instrument. The samples' pH was adjusted to the required value using stock solutions of HCl and NaOH. At least four measurements were taken for every data point, and the average was calculated.

2.5. Differential Scanning Calorimetry

Differential scanning calorimetry (DSC) was used to identify and characterize any thermal transitions of the constituents in the different samples during heating. Protein and mycoprotein dispersions (50–70 mg) were put into high volume aluminum pans, precisely weighed, and then hermetically sealed to avoid evaporation losses during the measurement process. Following equilibration to 25 °C, the sample cells were heated from 25 to 100 °C at a rate of 3 °C/min using the DSC apparatus (DSC25, TA Instruments, New Castle, DE, USA), as previously mentioned [30]. The heat flow (W/g) versus temperature (°C) profile of the samples was recorded as they were heated.

2.6. Dynamic Shear Rheology Analysis

Initially, PP, MCP and/or hybrid dispersions (pH 7.0, 100 mM NaCl) were prepared by dispersing the powdered ingredients in double distilled water and then stirring overnight at 350 rpm at room temperature. The rheological properties of these different samples were then analyzed under a dynamic shear rheometer (HR20, TA Instruments, New Castle, DE, USA). Around 1.0 to 1.5 mL of the unheated test sample was placed on a parallel plate (40.00 mm, stainless steel, crosshatched, Peltier plate) sample cell. The upper plate was then lowered onto the sample. The edges of the sample were then covered with mineral oil and a solvent trap was placed on top of the sample to reduce any evaporation during heating [34]. All analyses were performed using a gap of 1000 μm between the parallel plates.

- Temperature ramp: The samples were initially held at 25 °C for five minutes after being placed on the measuring cell. The rheological tests were then performed while the samples were heated at a rate of 6 °C per minute from 25 to 90 °C. After the samples reached 90 °C, they were maintained there for 20 min before being cooled to 25 °C at 6 °C/min. The samples were then kept at this temperature for another ten minutes. For these experiments, a frequency of 1.0 Hz and a strain of 0.1% were employed.
- Frequency sweep: After the temperature sweep was completed, a frequency sweep was performed, which involved measuring the dynamic shear modulus at 25 °C as a function of oscillation frequency (0.1–100 rad/s). A relatively small strain amplitude (0.1%) was employed during these tests to be within the linear viscoelastic region.
- Strain sweep: After the frequency sweep was completed, a strain sweep was carried out on the samples, which involved increasing the strain from 0.01 to 1000% at a fixed temperature of 25 °C and frequency of 1 Hz.

2.7. Texture Profile Analysis

Initially, powdered PP and/or MCP were dispersed in double-distilled water and then stirred overnight at 350 rpm at room temperature (pH 7.0, 100 mM NaCl). These dispersions were then poured into a glass flask, which was then heated for 30 min at 90 °C in a water bath before being cooled for two hours in an ice tray. The resulting gelled samples were then removed from the glass flasks and cut into cubes ($1 \times 1 \times 1$ cm^3). The textural characteristics of the gelled samples were then measured using an instrumental texture analyzer (TA-XT2, Stable Micro System, Surrey, UK) using single and double compression tests [35]:

- *Single compression test*: A single compression test was used to determine the Young's modulus and fracture characteristics of the samples, which was performed using a cylindrical probe (P/50, 50 mm stainless cylinder). The operating parameters used were based on those reported in a previous study: final target strain of 90%, and pre-test, test, and post-test speeds of 2, 1, and 10 mm/s, respectively [30]. The resultant stress–strain profiles were used to compute the Young's modulus, fracture stress, and fracture strain.
- *Double compression test*: A two-cycle compression/decompression program was used to ascertain the texture profile analysis (TPA) parameters of the samples. The fol-

lowing test parameters were used for these tests: final target strain was 50%, and pre-test, test, and post-test speeds were all set at 2 mm/s. The two different compression/decompression cycles were separated by 5 s. These experiments were also conducted using a cylindrical probe (P/50, 50 mm stainless cylinder) [30]. The generated force-distance profiles were used to calculate the samples' hardness, resilience, cohesion, springiness, gumminess, and chewiness, which was carried out by the computer software. The meaning of these terms, as well as criticisms of their limitations, have been given elsewhere [36].

2.8. Microstructure Analysis

CLSM: A confocal laser scanning microscope (CLSM) equipped with a 10× eye piece lens and a 40× objective lens (Nikon d-Eclipse C1 80i, Nikon, Melville, NY, USA) was used to determine the microstructure of the hybrid gels. Gel samples were placed on a glass slide after being cut into thin slices (0.2 mm) using a sharp knife. The samples were then stained using dyes specific to polysaccharides, lipids, and proteins: calcofluor white (1 mg/mL in ethanol) for polysaccharides, Nile red (1 mg/mL in water) for lipids, and FITC (1 mg/mL in ethanol) for proteins [31]. After adding one or two drops of these dyes to the samples, the mixtures were incubated for one to two minutes. A paper tissue was then used to remove any extra dye without affecting the sample. After that, the samples were examined under a microscope and covered with a glass cover slip. The instrument software was used to gather, store, and analyze the digital images (NIS-Elements 4.2, Nikon, Melville, NY, USA).

SEM: The microstructures of all the samples were also analyzed using a scanning electron microscope (SEM, JCM-6000 NeoScope, JEOL, Tokyo, Japan). The samples were first cut into 5 mm3 cubes and then dehydrated using a freeze drier for 3 days [37]. Following that, gold was sputter-coated onto the samples (Cressington 108Auto; Redding, CA, USA). The microstructure of the samples was then investigated under low vacuum conditions using an accelerating voltage of 10 kV.

2.9. Statistical Analysis

Apart from the dynamic shear rheology measurements, which were carried out in duplicate, all the other experiments were carried out in triplicate using newly prepared independent samples. Microsoft Excel (version 2405 (Build 17628.20144) was used to compute the means and standard deviations of this data. R-program software (R version 4.3.1) was used to execute an ANOVA (post hoc Tukey HSD test) to identify significant differences between samples ($p < 0.05$). It was assumed that the experimental data followed a normal distribution around the mean for the statistical analysis.

3. Results and Discussions

3.1. MCP Proximate Analysis

The proximate analysis of the MCPs showed that they contained 35.3–50.4% protein, 2.86% fat, 6.48 moisture, and 5.78% ash. This relatively wide range of protein contents is given because mycoproteins contain appreciable amounts of chitin, which contains nitrogen. Consequently, the nitrogen content (8%) measured by the Dumas method is influenced by both the protein and the chitin content of the mycoprotein. For this reason, the protein content was calculated based on conversion factors reported in previous studies for fungal proteins, which ranged from 4.38 [38] to 6.25 [39].

3.2. Electrical Characteristics

The pH of hybrid samples is important because it determines the electrical characteristics of the proteins and other components in the MCP and PP ingredients, which impacts their electrostatic interactions with each other. Previous studies have shown that electrostatic interactions play an important role in mycoprotein-plant protein hybrids [32,40]. For this reason, the zeta-potential values of the potato protein and mycoprotein dispersions were measured as a function of pH to provide some insights into their electrical characteristics (Figure 1). Potato proteins had an isoelectric point around pH 5, with the surface potential becoming increasingly negative at higher pH levels and increasingly positive at lower pH values. The change in zeta potential from positive, to zero, to negative as the pH was raised can be attributed to progressive conversion of the $-NH_3^+$ and $-COOH$ groups into $-NH_2$ and $-COO^-$ groups, respectively [41]. Around the isoelectric point, the electrostatic repulsion between the potato protein molecules is expected to be relatively weak and therefore insufficient to prevent their aggregation.

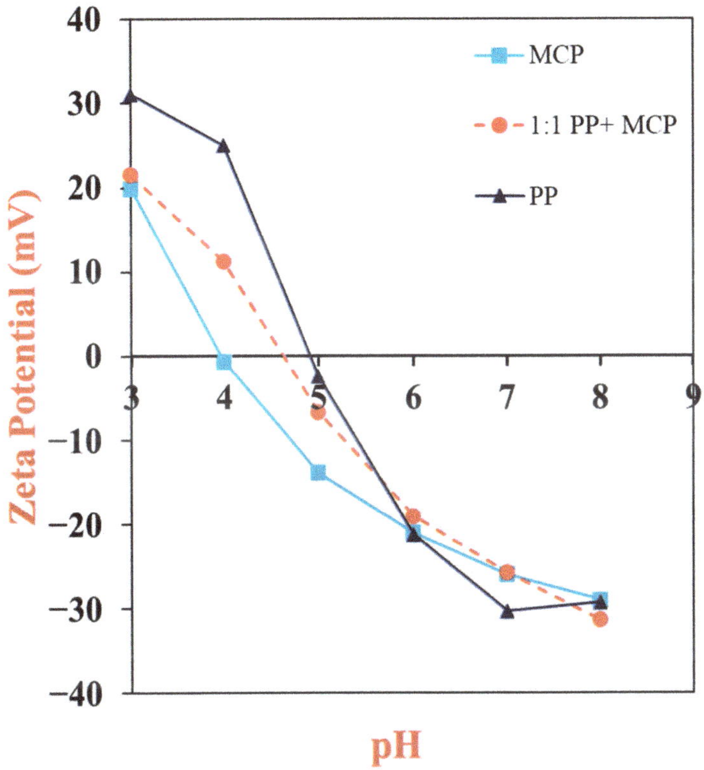

Figure 1. Impact of pH on the zeta potential of 0.1% (*w*/*v*) MCP, 0.1% (*w*/*v*) MCP + PP (1:1) and 0.1% (*w*/*v*) PP dispersions.

The zeta-potential versus pH profile of the mycoproteins followed a similar trend to the potato proteins. The charge changed from highly positive at low pH to highly negative at high pH, with an isoelectric point around pH 4. This isoelectric point agrees with that reported for fungal proteins in other studies [42]. Mycoproteins are known to be assembled from biopolymers that can carry charge, such as proteins and chitin [43]. The pH-dependence of the zeta potential of the mycoprotein dispersions is consistent with the presence of proteins at their surfaces, however, the amino groups on chitin may also have made some contribution to their positive charge at low pH values. This effect might partly account for the fact that the mycoproteins had a lower isoelectric point than the potato proteins.

The change in zeta potential with pH for the mixed PP + MCP system followed a similar pattern to the pure PP and pure MCP systems. Indeed, for most pH values the zeta-potential of the mixed system was between than of the two pure systems. The isoelectric point of the mixed system was around pH 4.7, which was between that of the pure PP (pH 4.9) and pure MCP (pH 4.0) systems. At pH values between the isoelectric points of the potato protein and mycoprotein systems (i.e., pH 4.0 to 4.9), the PP and MCP have opposite charges, which could promote their aggregation through increased electrostatic attraction.

In the remainder of this study, we only prepared samples at pH 7, since this is close to the pH of most meat products. In principle, however, it may be possible to create different structures and textures by varying the pH of the hybrid products, but this was beyond the scope of the current work.

3.3. Appearance of the Samples

Information about the pH-dependence of the aggregation of the different samples was obtained by taking digital photographs of them (Figure 2). The potato protein solutions were relatively clear at pH values far from their isoelectric point (pH 2, 3, 7, and 8), but turbid near their isoelectric point (especially pH 5), which is indicative of protein aggregation due to a reduction in electrostatic repulsion. The mycoprotein dispersions appeared turbid at all pH values, which can be attributed to light scattering by the small insoluble MCP fragments they contained. Moreover, a white precipitate formed at the bottom of these samples after a few hours of storage (Figure 2), which can be attributed to sedimentation of the MCP fragments caused by their relatively large size and higher density than water. The mixed MCP + PP dispersions contained precipitates at all pH values, which can mainly be attributed to the presence of the insoluble mycoprotein particles.

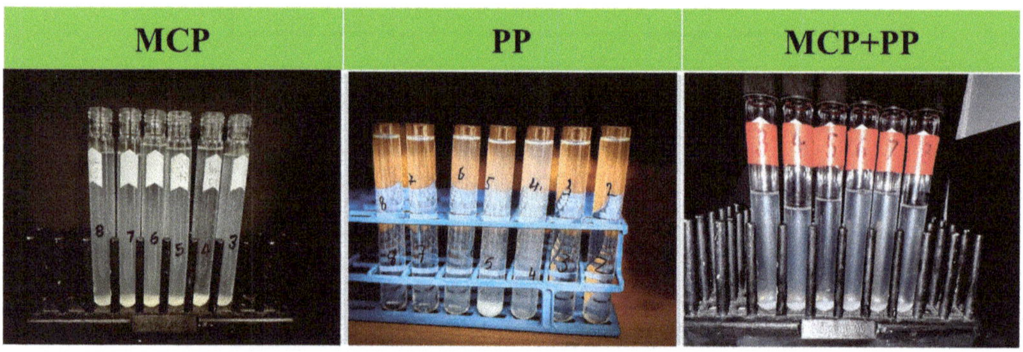

Figure 2. pH-dependence of the aggregation of the 0.1% (*w*/*v*) of MCP, 0.1% (*w*/*v*) PP and 0.1% (*w*/*v*) 1:1 MCP + PP samples.

3.4. Differential Scanning Calorimetry

The presence of any thermal transitions in the potato protein and mycoprotein dispersions was assessed using differential scanning calorimetry (DSC) analysis at pH 7. This pH is well above the isoelectric point of the potato proteins and mycoproteins, and so there should be a relatively strong electrostatic repulsion between them, which should reduce extensive protein aggregation [31]. The DSC thermograms for potato protein dispersions displayed a single, distinct endothermic peak upon heating, which is indicative of the thermal denaturation of the globular proteins [22,37]. The denaturation temperature (T_d) of the potato proteins exhibited a slight dependence on concentration, with a T_d of 65.80 °C and 65.53 °C at PP concentrations of 10% and 15%, respectively (Figure 3). This result agrees with that reported in our previous studies for potato proteins [31]. The MCP dispersions did not exhibit any thermal transitions during heating, which may have been because the proteins had already been denatured during the manufacturing processes used to prepare the powdered mycoproteins. Moreover, some of the proteins in the MCP are embedded within cell walls containing chitin, which may increase their thermal stability [44,45].

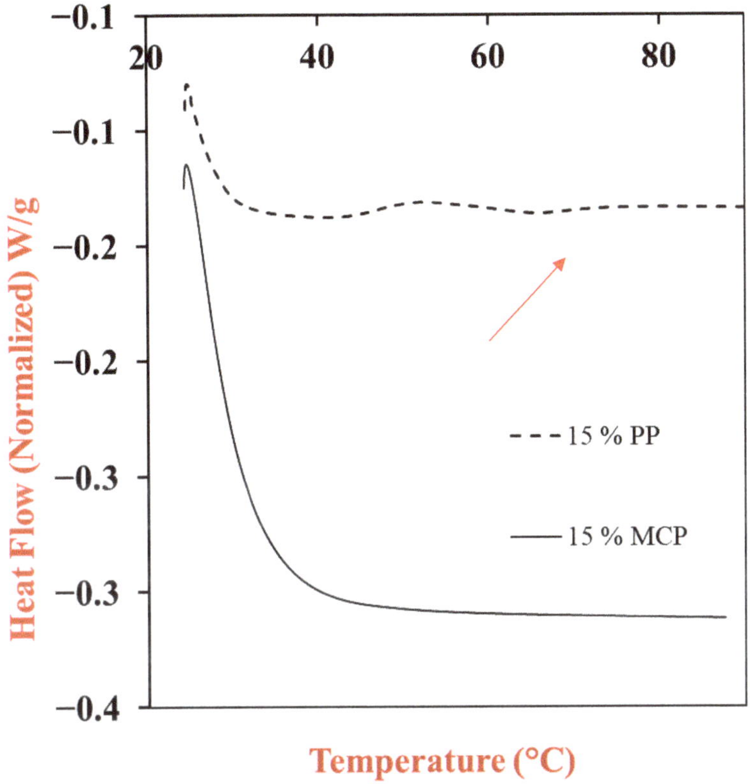

Figure 3. Differential scanning calorimetry profile of 15 wt% potato protein (PP) and 15 wt% mycoprotein (MCP) during heating from 25 to 100 °C with at 3 °C/min having denaturation temperature (T_d) of 65.53 °C indicated with arrow.

3.5. *Shear Rheology Analysis*

Dynamic shear rheology was used to provide information about the impact of hybrid composition on their rheological properties.

3.5.1. Temperature-Sweep Test

Initially, the temperature-dependence of the dynamic shear rheology of the potato protein and/or mycoprotein samples was determined. The storage modulus (G′) and loss modulus (G″) of these samples were measured as they were heated from 25 to 90 °C, held at this temperature, and then cooled to 25 °C. The rheology of 15% PP, 15% MCP, and 15% PP + 15% MCP hybrid samples are shown in Figure 4. Similar general trends were followed for samples containing 10% PP, 15% MCP, and 10% PP + 15% MCP but the shear modulus values were considerably lower. Notably, the samples containing 5% PP, 15% MCP or their hybrids did not form a gel or only formed a very weak gel that was easily disrupted.

For the samples containing 15% PP, the storage and loss moduli of the potato protein solutions were initially relatively low at 25 °C, which suggests that the proteins did not form a strong gel before heating (Figure 4a). This was probably because of the relatively strong electrostatic repulsion between the negatively charged protein molecules. When the temperature was raised, there was a slight decrease in the G′ and G″ values up to around 40 °C, which may have been due to dissociation of weak aggregates formed by the potato proteins at ambient temperature. Presumably, these aggregates formed due to the presence of some weak hydrophobic or hydrogen bonding between the protein molecules at lower temperatures. However, when the temperature was raised further, there was a modest increase in the shear moduli up to about 50 °C, followed by a steeper rise above about 60–70 °C, which can be ascribed to thermal denaturation and aggregation of the potato proteins leading to the formation of a 3D protein network with elastic-like properties [22]. Upon cooling from 90 to 25 °C, both G′ and G″ increased appreciably, indicating the formation of irreversible heat-set gels. The observed increase in gel strength during cooling can be attributed to strengthening of the hydrogen bonding within the protein matrix at lower temperatures [31].

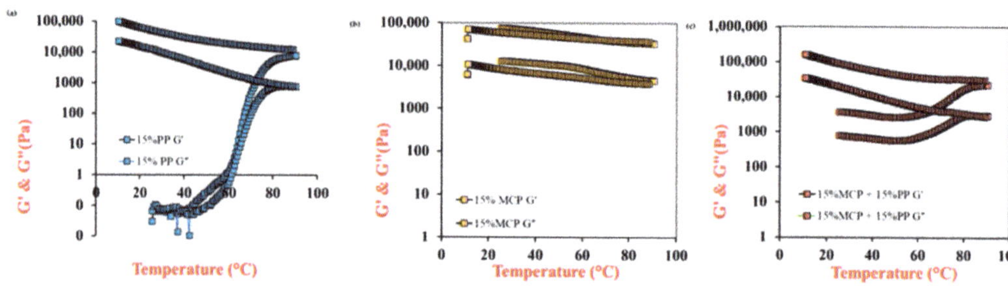

Figure 4. Temperature sweep results of (**a**) 15% potato protein (PP); (**b**) 15% mycoprotein (MCP); and (**c**) 15% MCP and 15% PP. The storage (G′) and loss (G″) moduli of the samples were measured as they were heated from 25 to 90 °C, held at 90 °C, and then cooled from 90 to 10 °C (strain = 0.1, frequency = 1 Hz). The red arrows show heating, while the blue arrows show cooling. For most temperatures, G′ > G″ for all samples, indicating they were predominantly elastic-like materials.

For the 15% MCP sample, both G′ and G″ were initially relatively high at 25 °C, which was consistent with the observed paste-like properties of these samples. There was a slight decrease in shear moduli when the samples were heated and a slight increase when they were cooled (Figure 4b). However, there was not a dramatic change in the shear rheology of the samples during either cooling or heating, suggesting there were no distinct thermal transitions, which is consistent with the DSC measurements (Figure 3). Notably, G′ remained higher than G″ during heating and cooling, indicating that they were predominantly elastic-like materials. Nevertheless, they were very soft solids, which were unsuitable for creating meat substitutes or analogs on their own. The reasons that the mycoproteins could not form strong gels may have been because the proteins were already denatured prior to heating or because they were trapped inside mycelium fibers and so they could not form a 3D network [40,46,47].

The mycoproteins are rich in proteins, dietary fibers, and micronutrients, and they naturally have a fibrous structure. These attributes are important to create meat analogs with good nutritional properties. However, the poor gelling properties of the mycoproteins limits their application for this purpose. Consequently, we examined the possibility of increasing the gel strength of mycoprotein-rich samples by blending them with potato proteins. The hybrid samples (15% PP + 15% MCP) initially had relatively high G′ and G″ values at 25 °C, which can mainly be attributed to the presence of the mycoproteins (Figure 4c). During heating, there was an initial softening of the hybrid hydrogels, which can be attributed to weakening of the hydrogen bonding between molecules at higher temperatures. However, when they were heated above about 60–70 °C there was a steep rise in gel strength, which can be attributed to unfolding and aggregation of the potato proteins. The gel strength then increased when the samples were cooled from 90 to 25 °C, which can be associated with strengthening of the hydrogen bonds between the biopolymers at lower temperatures. Notably, the final gel strength was higher for the hybrid samples than for the pure potato protein samples, which indicated that the addition of the mycoproteins strengthened the gels. It is possible that the fibers in the mycoproteins acted as active fillers, which are known to increase the elastic modulus of composite gels [30]. It should be noted that the final shear modulus (G′) of the hybrid samples after heating and cooling was quite similar to that reported for real cooked chicken under similar conditions [34]. Consequently, the hybrid products may have textural attributes that can match those of real meat products.

3.5.2. Frequency-Sweep Test

After the completion of the temperature-sweep test, a frequency-sweep test was performed to provide some insights into the dynamic properties of the hydrogels. For all the samples, there was a slight increase in the G′ and G″ values as the frequency was raised from 0.1 to 100 rad/s (Figure 5). This increase can be attributed to the fact that the biopolymers within the different hydrogels need a certain amount of time to rearrange themselves when an external shear stress is applied. At lower frequencies, the biopolymers have enough time to relax, but at higher frequencies, they do not, which leads to a more rigid and viscous material [37,48]. As expected, the magnitude of the shear modulus increased with increasing protein concentration in the PP samples. Moreover, the shear modulus of the hybrid samples was greater than that of the individual components with the corresponding PP concentration, which indicated that the presence of the mycoproteins again strengthened the potato protein gels. This effect may have been because some of the components within the MCP acted as active fillers that interacted with the surrounding protein network through attractive molecular interactions. For instance, MCP contains fibrous hyphae that are rich in β-glucan and chitin [43,49,50], which could form electrostatic, hydrophobic, and hydrogen bonds with the potato proteins [32].

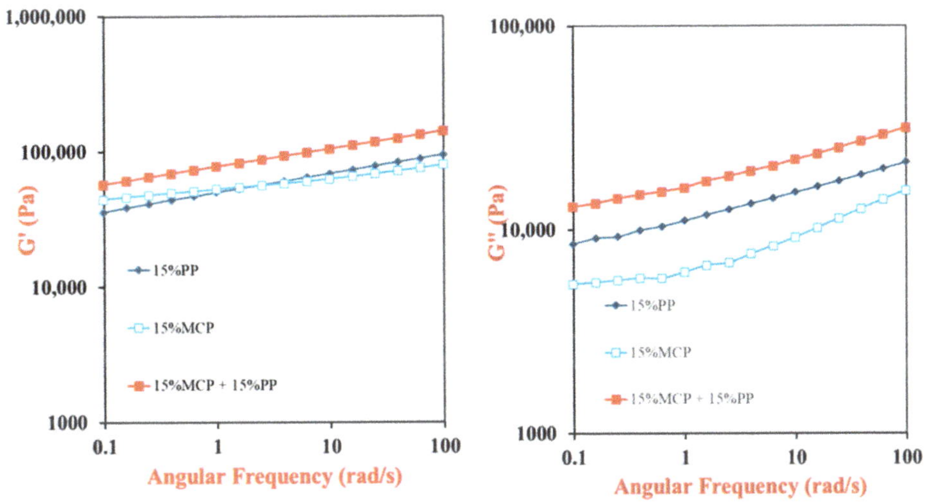

Figure 5. Frequency sweep results of 15% potato protein (PP), 15% mycoprotein (MCP), and 15% mycoprotein-15% potato protein (15% MCP + 15% PP) hybrid gels. The storage modulus (G′) and loss modulus (G″) of the samples were measured as the frequency was increased at a strain of 0.1% and 25 °C. For all frequencies, G′ > G″ for all samples, indicating they were predominantly elastic-like materials.

Some insights into the nature of the hydrogels formed can be obtained by analyzing the dependence of the shear modulus on frequency [31,51]. The experimental data can be fit to a power law model:

$$\log G' = z' \log \omega + K$$

Here, ω is the oscillation frequency, z' is a constant related to the frequency-dependence of the shear modulus, and K is a constant related to the strength of the molecular interactions. The constant z' represents the slope of a log-log plot of G' versus ω. Gels can be classified into different types depending on the z' value: covalent gels if $z' = 0$; physical gels if $z' > 0$; and viscous gels if $z' \leq 1$. The goodness of fit is evaluated from the coefficient of determination (R^2).

The parameters obtained by fitting the power-law model to the experimental data are reported in Table 1. In our experiments, R^2 was higher than 0.99 for all the samples, indicating a good fit of the model to the data. All our samples could be categorized as physical gels because the z' values were > 0. The most likely origin of the physical crosslinks between the biopolymer molecules in the different samples is hydrophobic, hydrogen, and electrostatic bonding [32,52]. The K values (6.74 to 11.26) of the gels depended on their composition, which suggests that the overall strength of the molecular interactions in them depended on the types of biopolymers present. Hydrophobic interactions will have occurred between non-polar groups on the surfaces of the proteins when they were heated above their thermal denaturation temperature. Hydrogen bonding may have occurred between polar groups on the proteins and/or polysaccharides present, especially at lower temperatures. Electrostatic interactions may have occurred between cationic chitin and anionic protein molecules in the hybrid samples. The importance of these interactions has been highlighted in other studies on mycelium–protein interactions. For example, researchers have used computer simulations, microscopy, rheology, and spectroscopy to provide insights into the nature of the molecular interactions involved in the creation of hybrid gels from egg white proteins and mycoproteins [45]. They reported that hydrogen bonding, electrostatic interactions, and hydrophobic interactions all played an important role. In another study, researchers used microscopy, rheology, and electrophoresis to provide insights into egg protein-mycoprotein interactions [40]. They concluded that electrostatic interactions played a dominant role, but that hydrogen bonding and hydrophobic interactions were also likely to be important. Similar findings have also been reported by the same research group for mycoprotein-potato protein interactions [32]. Taken together, the results of these studies support our hypothesis about the role of electrostatic, hydrophobic, and hydrogen bonding interactions in determining the properties of the hybrid gels.

Table 1. Impact of composition on the parameters determined by fitting a power-law model to the experimental shear modulus versus frequency data for single and hybrid gels.

	Storage Modulus			Loss Modulus		
Sample	**R^2**	**Z'**	**K'**	**R^2**	**Z''**	**K''**
5% PP	0.999	0.112	6.75	0.995	0.114	5.04
10% PP	0.999	0.147	8.73	0.997	0.137	7.84
15% PP	0.999	0.139	10.82	0.997	0.134	9.33
15% MCP + 5% PP	0.994	0.100	10.00	0.951	0.158	8.01
15% MCP + 10% PP	0.999	0.117	10.42	0.988	0.139	8.75
15% MCP + 15 PP	0.999	0.130	11.27	0.990	0.129	9.73
15% MCP	0.989	0.081	10.81	0.940	0.168	8.70

3.5.3. Strain-Sweep Test

After the completion of the frequency-sweep test, a strain-sweep test was carried out to provide some insights into their non-linear properties at high deformations. The viscoelastic properties of the different gels were assessed by measuring their complex shear modulus (G^*) when the strain was raised from 0.01% to 1000% (Figure 6).

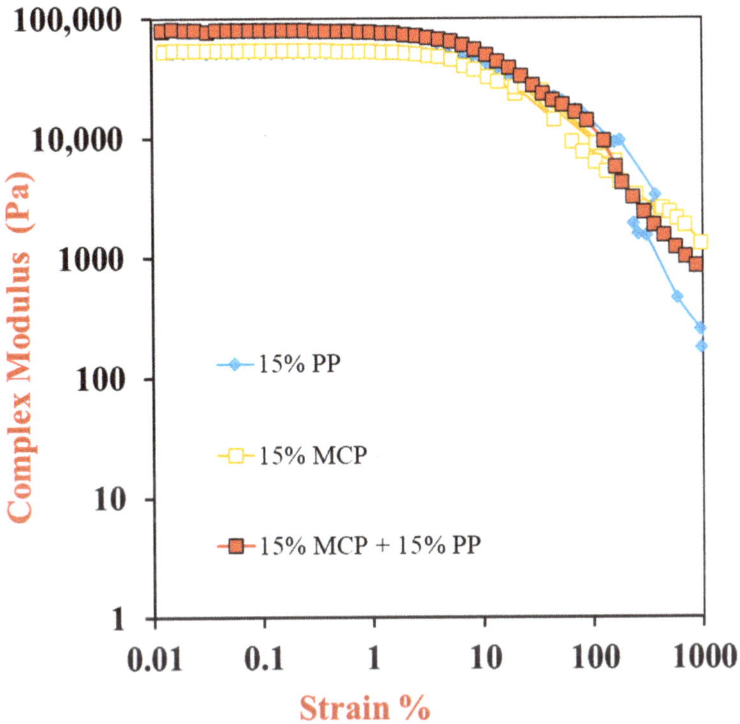

Figure 6. Strain sweep results of 15% pure potato protein,15% mycoprotein and15% potato protein-mycoprotein hybrid gels. The complex shear modulus (G*) of the samples were measured as the strain was increased from 0.01% to 1000% at 25 °C.

For all samples, the shear modulus remained relatively constant when the strain was increased up to about 2–4% but then it decreased steeply. This indicates that the different gels had reversible elastic properties at relatively low deformations, where the applied stress was proportional to the resulting strain. This low deformation region is referred to as the "linear viscoelastic regime". Above this region, the gel network is disrupted, and some yielding and flow may occur, which results in a decrease in the complex shear modulus. The strain sweep results therefore suggest that all the samples exhibited fairly similar breakdown profiles under shear. At higher strains, the applied shear stress was strong enough to disrupt the bonds between the different structural elements in the gel networks causing them to yield and flow.

3.6. Texture Profile Analysis

Further insights into the behavior of the different hydrogels when exposed to large deformations (like those experienced during mastication) were obtained using uniaxial compression testing [31]. For these tests, all the samples were heated to 90 °C for 30 min and then cooled to ambient temperature prior to analysis.

3.6.1. Single Compression Tests

A single compression test was used to measure the strain versus strain curves of the samples when they were compressed to a final strain of 90% (Figure 7). The Young's modulus and fracture properties of the gels were then determined (Table 2). The Young's modulus was calculated from the initial slope of the curves, whereas the breaking stress and breaking strain were determined from the first point when there was a break in these curves [35]. All the samples exhibited an initial rise in force with increasing strain, which indicated that they had some elastic-like properties. After reaching a certain strain threshold, all the samples then exhibited a notable increase in stress before ultimately failing, indicating their maximum loading capacity had been exceeded. The appreciable increase in force may have been because the surface area of the samples increased as they were compressed, and so they gave a greater resistance to compression. The samples may have broken at higher strains because the applied stresses exceeded the forces holding the different structures in the gel networks together. The breaking stress and strain values determined from the first observed break in the force-strain curves are reported in Table 2.

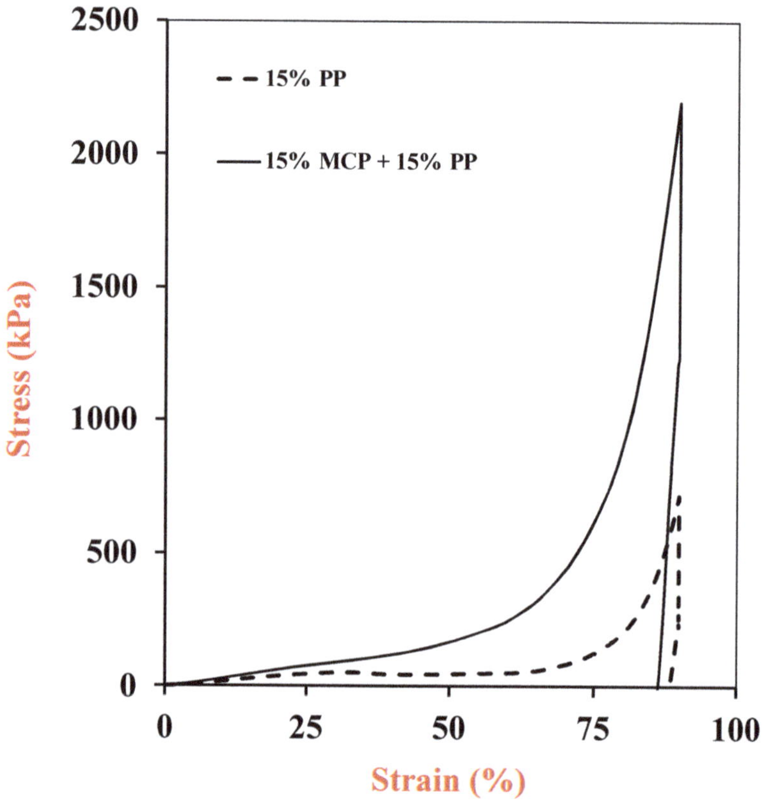

Figure 7. Effect of single composition on the stress–strain relationship of Potato protein and potato protein-mycelium hybrids during single compression-decompression experiments (25 °C).

Consistent with the shear modulus experiments, the Young's modulus of the 15% MCP + 15% PP hybrid gels was the highest. In addition, they also had the highest breaking stress (2143.5 kPa) and breaking strain (≥90.0%), which may be valuable for the formulation of meat analogs and substitutes. The enhanced mechanical properties of the hybrid gels may be attributed to a synergistic interaction between the MCP and PP components, which allows for a stronger and more robust biopolymer network to be formed. Indeed, fibrous fillers have been reported to be highly effective at increasing the mechanical strength of composite materials [53,54]. Notably, the Young's modulus of the hybrid samples (15% MCP + 15% PP) analyzed in our study had fairly similar values to that of cooked chicken reported in a previous study [34], which may be useful for their commercial application as meat substitutes.

Table 2. Impact of composition on Young's modulus, breaking stress, and breaking strain of single and hybrid gels measured using a single compression test. The 5% PP and 15% MCP samples were not analyzed (NA) because they were too soft and runny.

Sample	Young's Modulus (kPa)	Breaking Stress (kPa)	Strain (%)
5% PP	-	-	-
10% PP	0.520 ± 0.067	275 ± 16	≥90
15% PP	1.99 ± 0.23	567 ± 450	70 ± 35
15% MCP + 5% PP	0.141 ± 0.002	799 ± 73	≥90
15% MCP + 10% PP	0.323 ± 0.030	908 ± 28	≥90
15% MCP + 15% PP	2.71 ± 0.49	2144 ± 73	≥90
15% MCP	NA	NA	NA

Overall, the incorporation of the MCP into the PP gels enhanced their strength and flexibility, which may be partly due to the fibrous structures in the mycoproteins.

3.6.2. Double Compression Test

A double compression test was then used to measure the texture profile analysis (TPA) parameters of the different gels (Table 3). Representative force *versus* time profiles for different samples are shown in Figure 8, which show that incorporating the MCP into the potato protein gels increased their resistance to deformation. For the pure potato protein gels and the hybrid gels, the hardness and chewiness increased appreciably as the PP concentration was raised from 5 to 15% (Table 3). This effect can be attributed to the presence of a stronger protein network at higher protein concentrations due to greater crosslinking of the potato protein molecules. The increase in hardness with PP concentration is consistent with the increase in shear modulus and Young's modulus discussed earlier. The increase in chewiness with increasing potato protein content may be particularly important for the formulation of meat analogs and substitutes, as real meat products are usually characterized by a high degree of chewiness. There was not an appreciable dependence of the resilience (6.7–7.9%) or cohesiveness (0.23–0.33) of the pure potato protein samples on protein content. The resilience is a measure of how much the material regains its original shape and size after being compressed, while the cohesiveness is a measure of how well it can withstand a second deformation [36]. Our results suggest that all the pure potato protein samples exhibited relatively poor resilience and cohesiveness, which can be attributed to irreversible disruption of the 3D gel network formed by the proteins during compression.

Table 3. Impact of composition on the TPA parameters determined using a double compression test. The 5% PP and 15% MCP samples were not analyzed (NA) because they were too soft and runny.

Sample	Hardness (N)	Adhesiveness	Resilience (%)	Cohesion	Springiness (%)	Chewiness
5% PP	NA	NA	NA	NA	NA	NA
10% PP	1.7 ± 0.1	−0.18 ± 0.13	6.68 ± 0.98	0.33 ± 0.02	92.4 ± 4.9	0.51 ± 0.10
15% PP	6.2 ± 0.3	−0.12 ± 0.06	7.85 ± 0.53	0.23 ± 0.03	90.4 ± 4.3	1.26 ± 0.10
15% MCP + 5% PP	1.8 ± 0.4	−0.05 ± 0.03	12.46 ± 1.00	0.52 ± 0.03	83.2 ± 4.4	0.77 ± 0.21
15% MCP + 10% PP	4.4 ± 0.4	−0.03 ± 0.00	14.92 ± 0.22	0.59 ± 0.00	90.5 ± 1.4	2.35 ± 0.12
15% MCP + 15% PP	16.4 ± 1.7	−0.08 ± 0.02	13.05 ± 0.57	0.51 ± 0.01	84.1 ± 1.6	7.07 ± 0.71
15% MCP	NA	NA	NA	NA	NA	NA

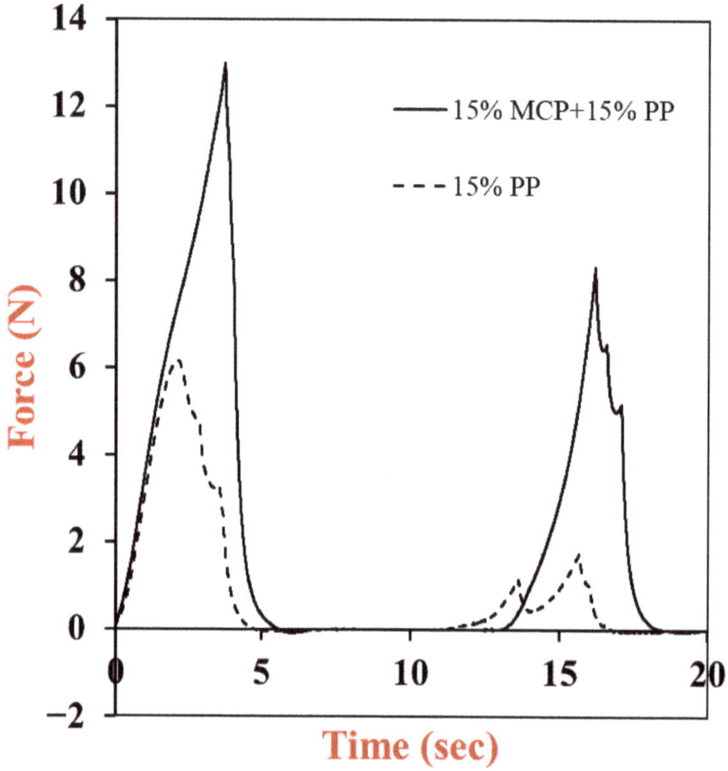

Figure 8. Double compression curves of 15% potato protein (15% PP) and 15% mycoprotein + 15% potato protein (15% MCP + 15% PP) hybrid gels. The force versus time curves were measured at 25 °C with 50% final strain and 2 mm/s pre-test, test, and post-test speeds.

The hardness and chewiness of the hybrid gels was greater than that of the equivalent potato protein gels. For example, the hardness of the 15% MCP + 15% PP gel was 16.4 ± 1.7 kPa, whereas that of the 15% PP gel was only 6.2 ± 0.3 kPa. Thus, the addition of the MCP increased the strength of the potato protein gels by nearly 165%, even though the MCP alone could not form a strong gel. As discussed earlier, this effect may have been because the MCP formed an interpenetrating network with the PP, or because some of the components in the MCP reinforced the molecular interactions between the potato protein molecules. There was not a strong dependence of the resilience (12.5–14.9%) or cohesiveness (0.51–0.59) of the hybrid gels on protein content. However, the resilience and cohesiveness values of the hybrid gels were significantly higher than those of the pure potato protein gels, which suggests that the presence of the mycoproteins somewhat increased their resistance to irreversible deformation during compression. This may have been because the fibrous hyphae in the MCP penetrated through the potato protein network, thereby holding it together better. The 5% PP and 15% MCP samples could not be analyzed using this method because they were too soft and runny.

3.7. Microstructure Analysis

Finally, the microstructures of the different heat-set gels were analyzed using scanning electron microscopy (SEM) and confocal laser scanning microscopy (CLSM) with fluorescence staining to provide some insights into the impact of the incorporation of the mycelium on the structure of the potato protein gels. It is often important to have a fibrous structure in meat analogs and substitutes. We hypothesized that the incorporation of the fibrous MCP into the PP gels would introduce this kind of structure in the hybrid gels. In addition, digital photographs of the different samples were taken to provide insights into the impact of their composition on their appearance. Only the 15% PP, 15% MCP, and 15% MCP + 15% PP hybrid gels were selected for the microstructure analysis because they had the strongest mechanical strength and were therefore the most suitable as meat substitutes and analogs.

3.7.1. CLSM Analysis

Confocal laser scanning microscopy with fluorescence staining was used to provide information about the microstructure of the different samples (Figure 9). In these images, the proteins were stained green, the polysaccharides blue, and the lipids red. The 15% mycoprotein gels had a complex interconnected network structure that consisted of fibrous proteins (green) and polysaccharides (blue). These fibers were probably fungal hyphae that consisted of chitin-rich walls with proteins packed inside [44]. There were some small yellowish-red regions in these images, which suggests that there were some lipids present, which is consistent with the proximate analysis of these samples (Table 1). There appeared to be many large pores (black regions) within these gels, which could affect the water-holding capacity, resilience, and texture of gels that include them. The fibrous structures found in the mycelia may be suitable for simulating those found in traditional meat products.

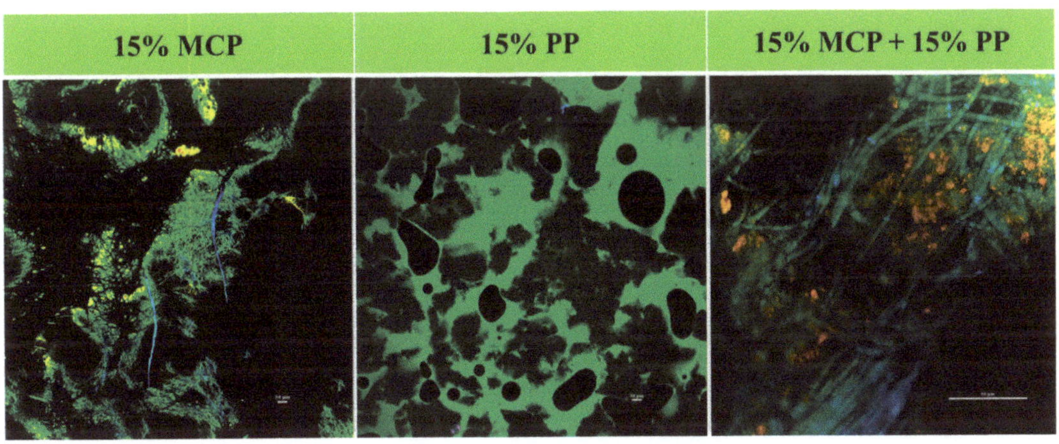

Figure 9. Confocal microscopy images of 15% mycoprotein (15% MCP), 15% potato protein (15% PP) and 15% mycoprotein + 15% potato protein (15% MCP + 15% PP) hybrid gels (25 °C). The images of the pure mycoprotein samples show they contained fibrous structures (stained blue and green), which were presumably chitin- and protein-rich hyphae. The images of the pure potato proteins showed that they contained large protein aggregates (stained dark green) dispersed in a protein-rich aqueous phase (stained light green). The black regions were probably holes formed during sample preparation. The images of the hybrid samples showed that they contained some fibrous structures (stained blue and green), which were presumably chitin- and protein-rich hyphae, distributed in a protein-rich network (stained green).

The 15% PP gels contained large dense irregular-shaped protein aggregates (dark green) dispersed within a dilute protein solution (light green). The large black holes seen in these images were voids in the gels that were introduced when they were cut into thin slices for microscopy analysis. These results are consistent with previous studies made on potato protein gels [52].

The CLSM images of the 15% mycoprotein and 15% potato protein hybrid gels showed that they had a highly heterogeneous network structure, with distinct regions corresponding to proteins (green), lipids (red), and polysaccharides (blue) (Figure 9). There appeared to be bundles of fibers from the mycelia wrapped around large protein-rich aggregates from the potato proteins. The complex microstructure of the hybrid gels would be expected to impact their functionality, including their water-holding capacity, resilience, and textural attributes [55].

3.7.2. SEM Analysis

Prior to SEM analysis, the gelled samples were freeze-dried at −80 °C to eliminate any moisture, which would interfere with the scanning electron microscopy measurements. Both low (300× magnification) and high (2000× magnification for 15% potato protein and 3000× for 15% mycoprotein and 15% mycoprotein + 15% potato protein hybrid gels) resolution images were taken of each of the samples to show their different levels of microstructure (Figure 10).

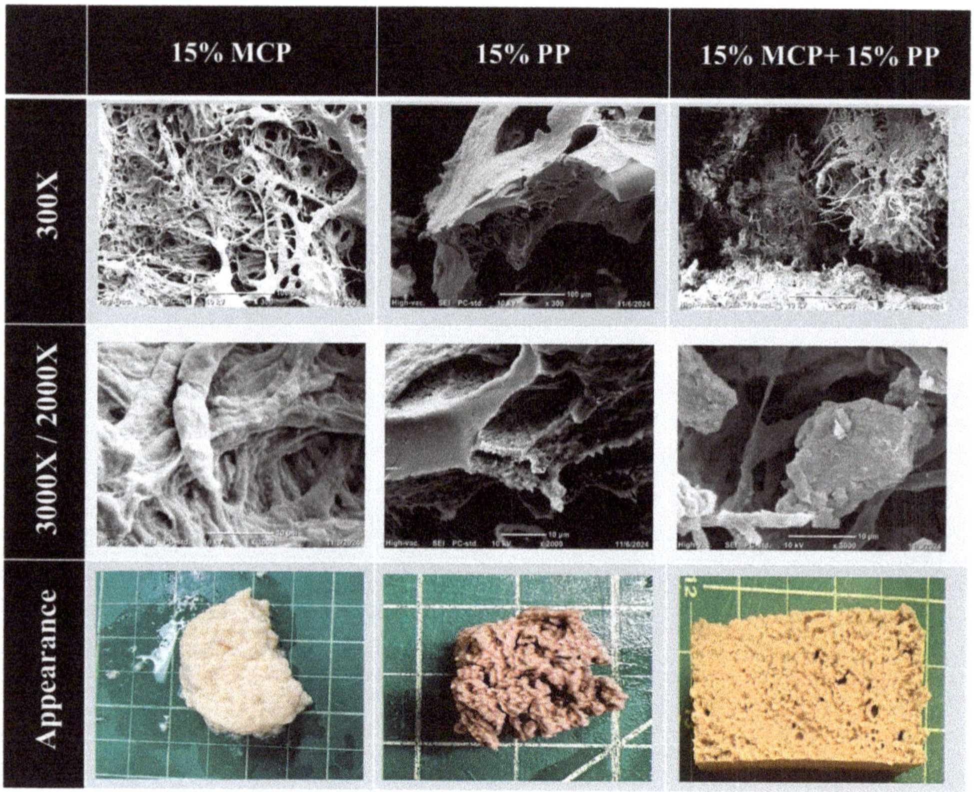

Figure 10. The scanning electron microscopy images of 15% mycoprotein, 15% potato protein and 15% mycoprotein + 15% potato protein hybrid gel. Scale bars are 100 μm for 300× and 10 μm for 3000× and 2000× for 15% potato protein. The digital photographs show the overall appearance of the samples before freeze drying.

The low-resolution images of the 15% PP samples showed they consisted of a relatively smooth matrix with some large irregular pores, which were probably air bubbles embedded within a uniform potato protein network. The high-resolution images of these samples indicated that the pores had rough surface textures, which were probably clusters of protein aggregates. These microstructural features are likely to play an important role in determining the mechanical strength and gelling behavior of the potato proteins.

The low-resolution images of the 15% MCP samples showed that they had a highly porous fibrous structure, but that the fibers did not seem to be aligned in any particular direction. This was probably because the original mycoprotein samples were ground into a find powder prior to use, which destroyed some of their original fibrous structures. The high-resolution images of these samples indicated that they contained an intricate network of fibrous structures. Overall, these images show that the mycoproteins have a highly fibrous microstructure, which may be useful for creating meat analogs or substitutes with meat-like textures.

The SEM images of the 15% MCP + 15% PP hybrid gels indicated that they had microstructural features that combined those seen in the MCP and PP samples. The low-resolution images show there were some regions that were relatively smooth, which may have been the potato protein network, as well as some regions that were fibrous, which may have been the mycoproteins.

The digital photographs of the samples showed that they had different appearances depending on their compositions. The 15% MCP samples were a whitish color and had a highly heterogeneous fibrous surface texture. The 15% PP samples had a beige color and a "chunky" appearance. The hybrid samples had an appearance that had some features of the two separate components, with a light yellowish-brown color and a fibrous porous texture.

In summary, the hybrid gels appeared to contain mycoprotein fibers embedded in a potato protein network. The mycoprotein provided a fibrous structure, whereas the potato proteins provided mechanical strength. These hybrids may therefore have some of the attributes required to create solid-like fibrous meat analogs and substitutes. However, sensory studies on meat analogs or substitutes created from these hybrid materials would be needed to confirm this.

4. Conclusions

In conclusion, this study showed that hybrid gels could be assembled by combining mycoproteins and potato proteins together. Each of these alternative protein sources brings different attributes to the hybrids. The potato proteins form a relatively strong gel network due to their ability to unfold and aggregate with each other during heating, leading to the formation of an irreversible heat-set gel. In contrast, the mycoproteins cannot form strong gels during heating, but they can provide a desirable fibrous structure, as well as valuable nutrients, such as proteins, dietary fibers, vitamins, and minerals. Consequently, these hybrid gels may be useful for formulating meat analogs and substitutes. The mechanical properties of these hybrid gels could be modulated by altering their protein concentration, with the gel strength increasing within potato protein concentration. The mechanical strength of the hybrid gels was greater than that of the equivalent potato protein gels, which suggested that the presence of the mycoproteins strengthened the gel network, which may have been because the mycelia fibers behaved as active fillers in the potato protein network.

The results of this study may be useful for the design and production of a new generation of plant protein-mycoprotein hybrids as sustainable food sources. However, further research is required to fully understand the molecular interactions driving their rheological and structural properties and to optimize their formulations for broader applications. In particular, it will be necessary to include other ingredients in commercial hybrid products, such as lipids, flavors, colors, and preservatives to make them more desirable to consumers. In addition, it will be important to examine the impact of other kinds of alternative protein sources, such as those from other plants, insects, or microbial fermentation, on the formation and properties of hybrid products formulated with mycoproteins. Moreover, it will be important to carry out sensory tests on these products to determine if they are acceptable to consumers. In principle, the fibrous structures and good nutritional profile provided by mycelia should reduce the number of ingredients and amount of processing steps required to formulate meat analogs, thereby increasing their consumer acceptance. Indeed, a recent study of the factors affecting consumer acceptance of fungal proteins in meat substitutes found that the sensory attributes of the final products were the most important factor, but that the number of ingredients and degree of processing were also important [56]. This study also found that many consumers were unfamiliar with fungal proteins and had a negative opinion of them because they associated them with mold. Another recent consumer study of the factors impacting the adoption of fungi-based foods reported that several factors were important, including sensory characteristics, environmental benefits, nutritional effects, production practices and ingredients. There is also a need to carry out more detailed life cycle analysis studies on hybrid products formed from mycelia, to establish their potential environmental impacts compared to other protein sources [17]. Moreover, there is a need to assess the economic viability of these products in terms of ingredient costs and processing methods. Consequently, further efforts are required to create higher quality hybrid foods using minimal processing methods, and on educating consumers about the potential health and environmental benefits of these sustainable food sources.

Author Contributions: Conceptualization, D.J.M.; Methodology, R.S. and D.J.; Formal analysis, R.S.; Investigation, R.S. and D.J.; Writing—original draft, R.S. and D.J.; Writing—review & editing, D.J.M.; Visualization, R.S.; Supervision, D.J.M.; Funding acquisition, D.J.M. All authors have read and agreed to the published version of the manuscript.

Funding: This material was partly based upon work supported by the National Institute of Food and Agriculture, USDA, Massachusetts Agricultural Experiment Station (MAS00559), and USDA, AFRI (2022-09185) grants.

Institutional Review Board Statement: Not applicable.

Informed Consent Statement: Not applicable.

Data Availability Statement: The original contributions presented in this study are included in the article. Further inquiries can be directed to the corresponding author.

Acknowledgments: The authors thank Sisheng Li, Jae Kun Ryu, and Xiaoyan Hu for valuable advice.

Conflicts of Interest: The authors declare no conflict of interest.

References

1. Bryant, C.J. Plant-based animal product alternatives are healthier and more environmentally sustainable than animal products. *Future Foods* **2022**, *6*, 100174. [CrossRef]
2. Poore, J.; Nemecek, T. Reducing food's environmental impacts through producers and consumers. *Science* **2018**, *360*, 987–992. [CrossRef]
3. Gerber, P.J.; Steinfeld, H.; Henderson, B.; Mottet, A.; Opio, C.; Dijkman, J.; Falcucci, A.; Tempio, G. *Tackling Climate Change Through Livestock: A Global Assessment of Emissions and Mitigation Opportunities*; CABI: Wallingford, UK, 2013.
4. Busch, J.; Ferretti-Gallon, K. What Drives and Stops Deforestation, Reforestation, and Forest Degradation? An Updated Meta-analysis. *Rev. Environ. Econ. Policy* **2023**, *17*, 217–250. [CrossRef]
5. Espinosa-Marrón, A.; Adams, K.; Sinno, L.; Cantu-Aldana, A.; Tamez, M.; Marrero, A.; Bhupathiraju, S.N.; Mattei, J. Environmental impact of animal-based food production and the feasibility of a shift toward sustainable plant-based diets in the United States. *Front. Sustain.* **2022**, *3*, 841106. [CrossRef]
6. Galanakis, C.M. The Future of Food. *Foods* **2024**, *13*, 506. [CrossRef]
7. Smetana, S.; Mathys, A.; Knoch, A.; Heinz, V. Meat alternatives: Life cycle assessment of most known meat substitutes. *Int. J. Life Cycle Assess.* **2015**, *20*, 1254–1267. [CrossRef]
8. Smetana, S.; Ristic, D.; Pleissner, D.; Tuomisto, H.L.; Parniakov, O.; Heinz, V. Meat substitutes: Resource demands and environmental footprints. *Resour. Conserv. Recycl.* **2023**, *190*, 106831. [CrossRef] [PubMed]
9. Springmann, M.; Clark, M.; Mason-D'Croz, D.; Wiebe, K.; Bodirsky, B.L.; Lassaletta, L.; De Vries, W.; Vermeulen, S.J.; Herrero, M.; Carlson, K.M. Options for keeping the food system within environmental limits. *Nature* **2018**, *562*, 519–525. [CrossRef] [PubMed]
10. Fernández-Rodríguez, R.; Bizzozero-Peroni, B.; Díaz-Goñi, V.; Garrido-Miguel, M.; Bertotti, G.; Roldán-Ruiz, A.; López-Moreno, M. Plant-based meat alternatives and cardiometabolic health: A systematic review and meta-analysis. *Am. J. Clin. Nutr.* **2024**. [CrossRef]
11. Alcorta, A.; Porta, A.; Tárrega, A.; Alvarez, M.D.; Vaquero, M.P. Foods for Plant-Based Diets: Challenges and Innovations. *Foods* **2021**, *10*, 293. [CrossRef] [PubMed]

12. McClements, D.J.; Grossmann, L. Next-Generation Plant-Based Foods: Challenges and Opportunities. *Annu. Rev. Food Sci. Technol.* **2024**, *15*, 79–101. [CrossRef]
13. Zhang, K.H.; Zang, M.W.; Wang, S.W.; Zhang, Z.Q.; Li, D.; Li, X.M. Development of meat analogs: Focus on the current status and challenges of regulatory legislation. *Compr. Rev. Food Sci. Food Saf.* **2023**, *22*, 1006–1029. [CrossRef]
14. Holt, R.R.; Munafo, J.P.; Salmen, J.; Keen, C.L.; Mistry, B.S.; Whiteley, J.M.; Schmitz, H.H. Mycelium: A Nutrient-Dense Food To Help Address World Hunger, Promote Health, and Support a Regenerative Food System. *J. Agric. Food Chem.* **2023**, *72*, 2697–2707. [CrossRef] [PubMed]
15. Majumder, R.; Miatur, S.; Saha, A.; Hossain, S. Mycoprotein: Production and nutritional aspects: A review. *Sustain. Food Technol.* **2024**, *2*, 81–91. [CrossRef]
16. Coelho, M.O.; Monteyne, A.J.; Dunlop, M.V.; Harris, H.C.; Morrison, D.J.; Stephens, F.B.; Wall, B.T. Mycoprotein as a possible alternative source of dietary protein to support muscle and metabolic health. *Nutr. Rev.* **2020**, *78*, 486–497. [CrossRef] [PubMed]
17. Mazac, R.; Järviö, N.; Tuomisto, H.L. Environmental and nutritional Life Cycle Assessment of novel foods in meals as transformative food for the future. *Sci. Total Environ.* **2023**, *876*, 162796. [CrossRef] [PubMed]
18. Dunlop, M.V.; Kilroe, S.P.; Bowtell, J.L.; Finnigan, T.J.; Salmon, D.L.; Wall, B.T. Mycoprotein represents a bioavailable and insulinotropic non-animal-derived dietary protein source: A dose–response study. *Br. J. Nutr.* **2017**, *118*, 673–685. [CrossRef]
19. McClements, D.J.; Grossmann, L. The science of plant-based foods: Constructing next-generation meat, fish, milk, and egg analogs. *Compr. Rev. Food Sci. Food Saf.* **2021**, *20*, 4049–4100. [CrossRef] [PubMed]
20. Prajapati, P.; Garg, M.; Singh, N.; Chopra, R.; Mittal, A.; Sabharwal, P.K. Transforming plant proteins into plant-based meat alternatives: Challenges and future scope. *Food Sci. Biotechnol.* **2024**, *33*, 3423–3443. [CrossRef] [PubMed]
21. van Vliet, S.; Kronberg, S.L.; Provenza, F.D. Plant-Based Meats, Human Health, and Climate Change. *Front. Sustain. Food Syst.* **2020**, *4*, 555088. [CrossRef]
22. Bhutto, R.A.; Bhutto, N.U.H.; Khanal, S.; Wang, M.W.; Iqbal, S.; Fan, Y.T.; Yi, J. Potato protein as an emerging high-quality: Source, extraction, purification, properties (functional, nutritional, physicochemical, and processing), applications, and challenges using potato protein. *Food Hydrocoll.* **2024**, *157*, 110415. [CrossRef]
23. Torres, M.D.; Dominguez, H. Valorisation of potato wastes. *Int. J. Food Sci. Technol.* **2020**, *55*, 2296–2304. [CrossRef]
24. Gorissen, S.H.M.; Crombag, J.J.R.; Senden, J.M.G.; Waterval, W.A.H.; Bierau, J.; Verdijk, L.B.; van Loon, L.J.C. Protein content and amino acid composition of commercially available plant-based protein isolates. *Amino Acids* **2018**, *50*, 1685–1695. [CrossRef]
25. Herreman, L.; Nommensen, P.; Pennings, B.; Laus, M.C. Comprehensive overview of the quality of plant- And animal-sourced proteins based on the digestible indispensable amino acid score. *Food Sci. Nutr.* **2020**, *8*, 5379–5391. [CrossRef]
26. Katzav, H.; Chirug, L.; Okun, Z.; Davidovich-Pinhas, M.; Shpigelman, A. Comparison of Thermal and High-Pressure Gelation of Potato Protein Isolates. *Foods* **2020**, *9*, 1041. [CrossRef]
27. Schmidt, J.M.; Damgaard, H.; Greve-Poulsen, M.; Sunds, A.V.; Larsen, L.B.; Hammershoj, M. Gel properties of potato protein and the isolated fractions of patatins and protease inhibitors—Impact of drying method, protein concentration, pH and ionic strength. *Food Hydrocoll.* **2019**, *96*, 246–258. [CrossRef]
28. Tiong, A.Y.J.; Crawford, S.; Jones, N.C.; McKinley, G.H.; Batchelor, W.; Hag, L.V. Pea and soy protein isolate fractal gels: The role of protein composition, structure and solubility on their gelation behaviour. *Food Struct.* **2024**, *40*, 100374. [CrossRef]
29. Kim, W.J.; Yiu, C.C.Y.; Wang, Y.; Zhou, W.B.; Selomulya, C. Toward Diverse Plant Proteins for Food Innovation. *Adv. Sci.* **2024**, *11*, e2408150. [CrossRef]
30. Santhapur, R.; Jayakumar, D.; McClements, D.J. Development and Characterization of Hybrid Meat Analogs from Whey Protein-Mushroom Composite Hydrogels. *Gels* **2024**, *10*, 446. [CrossRef] [PubMed]
31. Jayakumar, D.; Santhapur, R.; McClements, D.J. Preparation and characterization of plant protein-mushroom hybrids: Toward more healthy and sustainable foods. *Food Biophys.* **2024**, *19*, 1077–1094. [CrossRef]
32. Okeudo-Cogan, M.C.; Yang, S.Y.; Murray, B.S.; Ettelaie, R.; Connell, S.D.; Radford, S.; Micklethwaite, S.; Benitez-Alfonso, Y.; Yeshvekar, R.; Sarkar, A. Multivalent cations modulating microstructure and interactions of potato protein and fungal hyphae in a functional meat analogue. *Food Hydrocoll.* **2024**, *149*, 109569. [CrossRef]
33. Doane, T.L.; Chuang, C.H.; Hill, R.J.; Burda, C. Nanoparticle ζ-Potentials. *Acc. Chem. Res.* **2012**, *45*, 317–326. [CrossRef]
34. Ryu, J.; Xiang, X.; Hu, X.; Rosenfeld, S.E.; Qin, D.; Zhou, H.; McClements, D.J. Assembly of plant-based meat analogs using soft matter physics: A coacervation-shearing-gelation approach. *Food Hydrocoll.* **2023**, *142*, 108817. [CrossRef]
35. McClements, D.J.; Grossmann, L. *Next-Generation Plant-Based Foods: Design, Production, and Properties*; Springer Scientific: New York, NY, USA, 2022.
36. Peleg, M. The instrumental texture profile analysis revisited. *J. Texture Stud.* **2019**, *50*, 362–368. [CrossRef] [PubMed]
37. Ryu, J.; McClements, D.J. Control of plant-based biopolymer composite gel texture: Combining potato proteins with different high acyl-low acyl gellan gum ratios. *Food Hydrocoll.* **2024**, *149*, 109636. [CrossRef]
38. Kalač, P. Chemical composition and nutritional value of European species of wild growing mushrooms: A review. *Food Chem.* **2009**, *113*, 9–16. [CrossRef]
39. Bakratsas, G.; Polydera, A.; Nilson, O.; Kossatz, L.; Xiros, C.; Katapodis, P.; Stamatis, H. Single-cell protein production by Pleurotus ostreatus in submerged fermentation. *Sustain. Food Technol.* **2023**, *1*, 377–389. [CrossRef]

40. Okeudo-Cogan, M.C.; Murray, B.S.; Ettelaie, R.; Connell, S.D.; Radford, S.J.; Micklethwaite, S.; Sarkar, A. Understanding the microstructure of a functional meat analogue: Demystifying interactions between fungal hyphae and egg white protein. *Food Hydrocoll.* **2023**, *140*, 108606. [CrossRef]
41. Ryu, J.; McClements, D.J. Cellulose reinforcement of plant-based protein hydrogels: Effects of cellulose nanofibers and nanocrystals on physicochemical properties. *Food Hydrocoll.* **2025**, *158*, 110541. [CrossRef]
42. Wang, J.; Tian, Q.; Cui, L.; Cheng, J.; Zhou, H.; Zhang, Y.; Peng, A.; Shen, L. Synergism and mutualistic interactions between microalgae and fungi in fungi-microalgae symbiotic system. *Bioresour. Technol.* **2022**, *361*, 127728. [CrossRef]
43. Colosimo, R.; Warren, F.J.; Edwards, C.H.; Ryden, P.; Dyer, P.S.; Finnigan, T.J.A.; Wilde, P.J. Comparison of the behavior of fungal and plant cell wall during gastrointestinal digestion and resulting health effects: A review. *Trends Food Sci. Technol.* **2021**, *110*, 132–141. [CrossRef]
44. Fernando, L.D.; Dickwella Widanage, M.C.; Penfield, J.; Lipton, A.S.; Washton, N.; Latgé, J.-P.; Wang, P.; Zhang, L.; Wang, T. Structural polymorphism of chitin and chitosan in fungal cell walls from solid-state NMR and principal component analysis. *Front. Mol. Biosci.* **2021**, *8*, 727053. [CrossRef] [PubMed]
45. Gow, N.A.; Latge, J.-P.; Munro, C.A. The fungal cell wall: Structure, biosynthesis, and function. *Microbiol. Spectr.* **2017**, *5*. [CrossRef]
46. Muhedaner, M.; Bako, H.K.; Zhou, G.; Ye, K. Impact of egg white protein on mycoprotein gel: Insights into rheological properties, protein structure and Molecular interactions. *Food Chem.* **2025**, *463*, 141366. [CrossRef] [PubMed]
47. Chulikavit, N.; Huynh, T.; Dekiwadia, C.; Khatibi, A.; Mouritz, A.; Kandare, E. Influence of growth rates, microstructural properties and biochemical composition on the thermal stability of mycelia fungi. *Sci. Rep.* **2022**, *12*, 15105. [CrossRef]
48. Hu, X.; Ju, Q.; Koo, C.K.; McClements, D.J. Influence of complex coacervation on the structure and texture of plant-based protein-polysaccharide composites. *Food Hydrocoll.* **2024**, *147*, 109333. [CrossRef]
49. Colosimo, R.; Harris, H.C.; Ahn-Jarvis, J.; Troncoso-Rey, P.; Finnigan, T.J.A.; Wilde, P.J.; Warren, F.J. Colonic in vitro fermentation of mycoprotein promotes shifts in gut microbiota, with enrichment of *Bacteroides* species. *Commun. Biol.* **2024**, *7*, 272. [CrossRef]
50. Colosimo, R.; Warren, F.J.; Finnigan, T.J.A.; Wilde, P.J. Protein bioaccessibility from mycoprotein hyphal structure: In vitro investigation of underlying mechanisms. *Food Chem.* **2020**, *330*, 127252. [CrossRef] [PubMed]
51. Egelandsdal, B.; Fretheim, K.; Harbitz, O. Dynamic rheological measurements on heat-induced myosin gels: An evaluation of the method's suitability for the filamentous gels. *J. Sci. Food Agric.* **1986**, *37*, 944–954. [CrossRef]
52. Ryu, J.; McClements, D.J. Impact of heat-set and cold-set gelling polysaccharides on potato protein gelation: Gellan gum, agar, and methylcellulose. *Food Hydrocoll.* **2024**, *149*, 109535.
53. Ahmad, Z.; Mark, J.E. Biomimetic materials: Recent developments in organic-inorganic hybrids. *Mater. Sci. Eng. C-Biomim. Supramol. Syst.* **1998**, *6*, 183–196. [CrossRef]
54. Nitai, A.S.; Chowdhury, T.; Inam, M.N.; Rahman, M.S.; Mondal, M.I.H.; Johir, M.A.H.; Hessel, V.; Fattah, I.M.R.; Kalam, M.A.; Suwaileh, W.A.; et al. Carbon fiber and carbon fiber composites-creating defects for superior material properties. *Adv. Compos. Hybrid Mater.* **2024**, *7*, 169. [CrossRef]
55. Zhang, H.; Wu, J.; Cheng, Y. Mechanical properties, microstructure, and in vitro digestion of transglutaminase-crosslinked whey protein and potato protein hydrolysate composite gels. *Foods* **2023**, *12*, 2040. [CrossRef] [PubMed]
56. Chezan, D.; Flannery, O.; Patel, A. Factors affecting consumer attitudes to fungi-based protein: A pilot study. *Appetite* **2022**, *175*, 106043. [CrossRef]

foods

MDPI

Article

Unravelling the Digestibility and Structure–Function Relationship of Lentil Protein Through Germination and Molecular Weight Fractionation

Armin Mirzapour-Kouhdasht [1,2], Samaneh Shaghaghian [3,4], Marjan Majdinasab [3], Jen-Yi Huang [2] and Marco Garcia-Vaquero [1,*]

1 Section of Food and Nutrition, School of Agriculture and Food Science, University College Dublin, Belfield, Dublin 4, Ireland; amirzapo@purdue.edu
2 Department of Food Science, Purdue University, West Lafayette, IN 47907, USA; huang874@purdue.edu
3 Department of Food Science and Technology, School of Agriculture, Shiraz University, Shiraz 71441-65186, Iran; samaneh.shaghaghian.1@ulaval.ca (S.S.); majdinasab@shirazu.ac.ir (M.M.)
4 Department of Food Science, University of Laval, Quebec, QC G1V0A6, Canada
* Correspondence: marco.garciavaquero@ucd.ie

Abstract: This study explores for the first time the impact of a 6-day germination process on the structure (FTIR), antioxidant activity, nutritional/safety attributes (ACE-I inhibitory activity, digestibility, and cytotoxicity), and functional properties of fractions of variable molecular weight (W > 5 kDa; 3 kDa < MW < 5 kDa; and MW < 3 kDa) isolated from proteins extracted from lentils. FTIR results indicated a substantial increase in β-sheet contents during germination. The digestibility of proteins increased from day 0 (16.32–17.04%) to day 6 of germination (24.92–26.05%) with variable levels of digestibility depending on their MW. ACE-I inhibitory activity improved during germination in all fractions, reaching IC_{50} values of 0.95, 0.83, and 0.69 mg/mL after 6 days of germination. All antioxidant activities analyzed notably increased, particularly in low-MW fractions (MW < 3 kDa). The functional properties of low-MW fractions were also the most promising, displaying the highest water and fat binding capacities and emulsifying and foaming capacities but lower foaming and emulsifying stability compared to high-MW fractions. Cytotoxicity tests on L929 cells revealed the slight adverse effects of low-MW fractions during germination. This study provides insights into the enhanced nutritional and functional attributes of lentil proteins following germination, emphasizing their potential application in functional foods.

Keywords: lentil seeds; protein digestibility; ACE-I inhibitory activity; antioxidant; germination

1. Introduction

Lentils (*Lens culinaris*) are a pulse crop belonging to the *Leguminosae* family and are one of the most highly consumed pulses around the globe [1]. The consumption of lentils has been linked to multiple nutritional benefits, including cholesterol- and lipid-lowering effects, and reducing the risk of cancer and type-2 diabetes, facts that could be attributed to their high dietary fiber and phenolic contents and to the high antioxidant capacity of the compounds produced by this crop [2].

The relatively high protein content of lentils, ranging from 24 to 30% (*w*/*w*), has recently attracted the attention of the food industry aiming to use lentil protein flours to fortify with plant protein novel food formulations [1,3]. Previous studies have demonstrated that among different pulses, lentils also contain high-quality protein and that its consumption is linked to antihypertensive effects [4]. Lentils are considered a rich source of isoleucine, leucine, lysine, threonine, phenylalanine, and valine compared to other crops while being deficient in methionine and cysteine [5]. The large amount of phenylalanine in lentils has been linked to an increased production of tyrosine which is needed for the formation of epinephrine and norepinephrine, contributing to improving neurotransmission activity under stress conditions [5,6]. Whereas a healthy adult synthesizes tyrosine from phenylalanine, a young child has not yet developed the enzyme phenylalanine hydroxylase needed for this conversion. Thus, for this sector of the population, tyrosine is an essential amino acid. This phenomenon can also occur in various pathological conditions [7]. In addition, branched-chain amino acids (BCAAs: leucine, valine, and isoleucine) are particularly important in older adults. BCAAs have a physiological role during protein synthesis, metabolism, food intake, and aging. Many studies have contradictory conclusions concerning the relationship between the blood levels of BCAAs or the dietary manipulation of these amino acids and changes appreciated during aging in body composition, sarcopenia, obesity, insulin, and glucose metabolism [8].

Lentil proteins are generally extracted by conventional methods that include a two-stage process, alkaline extraction (pH 8–11) followed by isoelectric precipitation (pH 4–5) [6]. Under these conditions, approximately 70% (*w*/*w*) of lentil protein is composed of globulins or salt-soluble protein [9] that can be further classified as vicilin (7S), with a molecular weight (MW) of 40–70 kDa and legumin (11S) with a MW of 320–380 kDa. The ratio of 7S/11S can change depending on the lentil variety under study; however, it is regularly described as being around 2.78 [1,9].

Despite all the aforementioned benefits of lentil seeds and their derived protein isolates, lentils, similarly to other pulses, may contain some anti-nutritional compounds (ANCs), such as tannins, lectins, phytic acid, and protease inhibitors, which may reduce their nutritional benefits. However, the process of protein extraction (alkali followed by acid) has been reported to alleviate the adverse effects of ANCs, improving the nutritional value and digestibility of lentil proteins [9]. Moreover, germination is a traditional process and a re-emerging trend in the food industry to unlock the nutritional potential of seeds by making them more readily available for human consumption [10,11]. Thereby, S. Santos, Silva, MP Valente, Gruber, and W. Vasconcelos [4] reported a significant increase, of around 30%, in protein content in different lentil varieties after sprouting.

Multiple enzymes are activated during the process of the germination of seeds, contributing to the transformation of complex molecules (carbohydrates, proteins, and fats) into small forms that are easier to digest and absorb by the human body; thus, variations in the molecular weight (MW) and properties of the protein ingredients achieved from these seeds are likely to change during the process of germination. Germination has been previously reported to increase the bioavailability of nutrients, such as vitamins, minerals, amino acids, and antioxidants, present in seeds [10,11]. From a food technology perspective, germination can be used to enhance the functional properties of the ingredients derived from these seeds. Germinated seeds can be used in bread-making to increase the protein quality and improve the texture of the final baked products. The seeds can also be sprouted and used as ingredients in plant-based milk alternatives or energy bars [10,11]. Aligned with this concept, the germination of lentils has been described as having positive changes in nutrient availability and the digestibility of the seeds, reducing anti-nutrients, and increasing bioactive compounds and the nutritional value of their proteins [12,13], making them an interesting ingredient for the development of functional foods.

This study investigated the nutritional, physicochemical, and functional changes in lentil proteins extracted from seeds during a 6-day germination period. The proximate composition of lentils (moisture, protein, lipid, carbohydrate, and fiber contents) and changes in their structure—measured by Fourier transform infrared (FTIR) analysis—were determined during the 6-day germination period. Moreover, fractions of variable MW were generated from all the protein isolates during all stages of germination to examine their digestibility, antioxidant activity (DPPH radical scavenging, ferrous chelating activity, hydroxyl radical scavenging activity, and ferric reducing antioxidant power), cytotoxicity (L929 cells), and functional properties (water/oil binding capacity and emulsifying and foaming properties) to provide a complete perspective of the changes induced by lentil proteins during the process of germination that could have an impact for the development of value-added products from different seed sprouts in the food industry.

2. Materials and Methods

2.1. Chemical Reagents

Pepsin from porcine stomach mucosa pepsin (3200 U/mg) (SRE0001) and pancreatin from porcine pancreas (P1750) were purchased from Merck (Arklow, County Wicklow, Ireland). The angiotensin-converting enzyme-I (ACE-I) inhibition assay kit-WST was purchased from Dojindo Laboratories (Kumamoto, Japan). L929 cells (mouse fibroblast, subcutaneous connective tissue) were obtained from the European Collection of Animal Cell Cultures (Salisbury, UK). All other chemicals and reagents were of analytical grade.

2.2. Seeds and Germination Process

Lentil seeds (*Lens esculenta*) were purchased from a local market in Shiraz, Iran. All seeds were washed with distilled water to remove impurities. The germination process was performed following the procedure described by Fouad and Rehab (2015) [14] with slight modifications. The lentil seeds were soaked in distilled water at a ratio of 1:10 (*w*/*v*) overnight. The water from this soaking was then discarded, and the seeds were wrapped using a wet cotton cloth and kept in dark conditions. The seeds were irrigated daily with fresh distilled water for 6 days and the cotyledon emergence dynamics and changes during that period can be appreciated in Figure 1. Samples were selected at 1-day intervals and frozen at −20 °C to cease germination. After thawing, the samples were oven-dried (50 °C, 6 h) and grinded (Kyocera, CM-20C SF, Kyoto, Japan). All samples were vacuum packed and stored at −20 °C for further analyses according to the experimental plan detailed in Figure 2.

Figure 1. The evolution of lentil seeds during the germination process over a 6-day period.

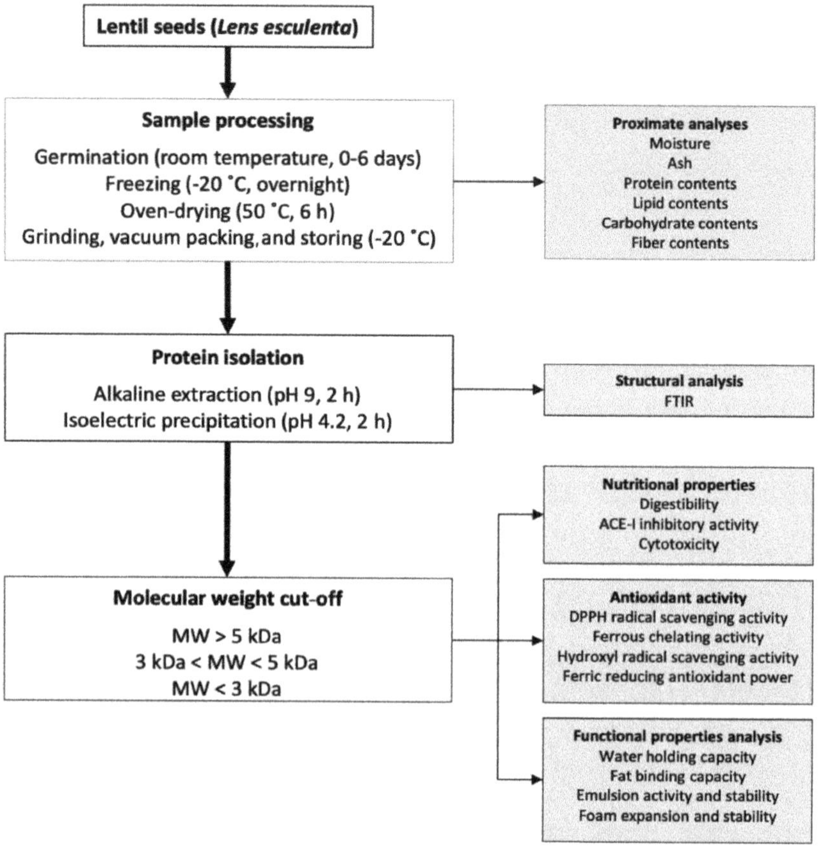

Figure 2. A schematic flow diagram illustrating the experimental design followed during this study.

2.3. Proximate Analyses

Moisture and ash contents were determined using the official method of analysis, AOAC.942.05, as described by Horowitz (2000) [15]. The protein content of the samples was assessed by the Kjeldahl method (BUCHI Labortechnik AG, Flawil, Switzerland) using a nitrogen-to-protein conversion coefficient of 6.25 [16]. Lipid contents were measured using Soxhlet apparatus, following the procedure outlined by Connolly, Piggott, and FitzGerald (2013). Briefly, 1 g of the samples was mixed with 100 mL of hexane (68 °C, 6 h) and the solvent evaporated in a rotary evaporator (Stuart, RE401, UK). The carbohydrate content of the samples was determined following the method as described by Masuko et al. (2005) [17]. Briefly, 30 μL of the samples or a standard (glucose, 10–50 mg/mL) was mixed with 150 μL of concentrated sulfuric acid (98.5%) and 30 μL of 5% phenol solution and shaken for 2 min. The mixtures were incubated in a water bath (90 °C, 5 min), cooled down at room temperature for 10 min, and agitated for 10 min before the absorbance of the reactions was read at 490 nm using a spectrophotometer (BioTek, Winooski, VT, USA).

The fiber content of the samples was evaluated as outlined by Lee, Prosky, and Vries (1992) [18]. Briefly, 1 g of the samples was mixed with 40 mL of maleate buffer (50 Mm, pH 6.0) containing 2 mM $CaCl_2$, 50 U/mL porcine pancreatic α-amylase, and 3.4 U/mL amyloglucosidase, and the solutions were incubated (37 °C, 16 h). The pH of the mixtures was adjusted to 8.2 before adding 0.1 mL of *Bacillus licheniformis*-derived protease (350 U/mL), and the samples were incubated (60 °C, 30 min), cooled down to room temperature, and mixed with 4 volumes of 95% ethanol, and the filtrates were oven-dried (105 °C, overnight). The fiber contents were calculated using Equation (1).

$$\text{Fiber content } (\%) = \frac{W1 - W2}{W1} \times 100 \quad (1)$$

W_1 and W_2 are the initial weight of the sample and weight of the dried filtrate, respectively.

2.4. Protein Isolation Process

Proteins were isolated from the powdered samples as described by Suliman, El Tinay, Elkhalifa, Babiker, and Elkhalil (2006) [19] with slight modifications. Briefly, 30 g of the samples was suspended in 500 mL of distilled water, and their pH was adjusted to 9 using 1 M NaOH. After mixing for 2 h, the samples were centrifuged (3000× *g*, 30 min). The supernatants were collected and the pH was adjusted to 4.2 using 1 M HCl. The solutions were mixed for 2 h before centrifugation (10,000× *g*, 20 min); the precipitated proteins were rehydrated, neutralized to a pH of 7, freeze-dried, and stored at −20 °C for further experiments.

2.5. Molecular Weight Cut-Off Fractionation of Protein Isolates

Protein isolates were re-dissolved in deionized water (1:10 *w*/*v*) and sequentially filtered through 3 and 5 kDa centrifugal ultrafilter tubes (Millipore, Sigma-Aldrich, St. Louis, MO, USA). The process resulted in 3 different MWCO fractions of variable MW: (1) MW > 5 kDa; (2) 3 kDa < MW < 5 kDa; and (3) MW < 3 kDa. All the fractions were freeze-dried and stored at −20 °C for further experiments.

2.6. Fourier Transform Infrared (FTIR) Analysis

The secondary structural compositions of germinated and ungerminated lentil protein isolates were analyzed using a Fourier transform infrared (FTIR) spectrometer equipped with an attenuated total reflectance cell (Tensor II, Bruker, Germany). The results were recorded three times, using a wavelength range of 4000 to 400 cm^{-1} (mid-infrared region) for each sample. The second-derivative analysis was conducted using OriginPro software (version 9.0, 2013) computing the areas of different spectral components, showing the relative proportion of secondary structures.

2.7. Nutritional Properties

2.7.1. Digestibility

The digestibility of samples was assessed following the procedure outlined by Marrion et al. (2005) [20] with slight modifications. Samples, with an approximately 40 mg nitrogen content, were digested with pepsin (1:60,000, 30 min, pH 2, and 37 °C). Subsequently, the samples were treated with pancreatin (pH 7.5, 37 °C) inside dialysis bags for 6 h. The digestibility of the samples was calculated using Equation (2).

$$\text{Digestibility } (\%) = \frac{N \text{ } in \text{ } dialysates \text{ } (mg)}{N \text{ } in \text{ } sample \text{ } (40 \text{ } mg)} \times 100 \quad (2)$$

2.7.2. Angiotensin-Converting Enzyme I (ACE-I) Inhibitory Activity

The ACE-I inhibitory activity of all samples was tested using the ACE-I inhibition kit-WST (Dojindo EU GmbH, Munich, Germany) following the manufacturers' recommendations. Briefly, after preparing enzyme B solution (by dissolving it in 2 mL of deionized water), 1.5 mL of enzyme B solution was added to enzyme A, generating the enzyme working solution. Enzyme C and coenzyme were both dissolved in 3 mL of deionized water and 2.8 mL of each solution was combined, producing the indicator working solution.

Blank 1 was established by adding 20 μL of deionized water and 20 μL of substrate buffer. Blank 2 was created by adding 40 μL of deionized water and 20 μL of substrate buffer. The inhibitors were tested by adding 20 μL of the unknown samples (0, 0.25, 0.5, 0.75, 1, 1.5, and 2 mg/mL) or positive control (captopril solution, 15 ng/mL) with 20 μL of substrate buffer. Afterwards, 20 μL of enzyme working solution was added to each inhibitor and negative control well to initiate the enzymatic reaction. The microplates were incubated (dark conditions, 37 °C, 1 h) under constant agitation (50 rpm). Following incubation, 200 μL of indicator working solution was added to each well and the plates were incubated for 10 min at room temperature. The absorbance of the samples was read at 450 nm using a spectrophotometer (BioTek, Winooski, VT, USA). The ACE-I inhibitory activity of the samples was calculated using the following equation (Equation (3)).

$$\text{ACE} - \text{I inhibition } (\%) = \frac{(\text{AB1} - \text{AI})}{\text{AB1} - \text{AB2}} \times 100 \quad (3)$$

where A_{B1}, A_I, and A_{B2} are the absorbance of the control, sample, and blank, respectively.

The half-maximal inhibitory concentration (IC_{50}) was considered the concentration of the sample at which 50% of ACE-I activity was inhibited.

2.7.3. Cytotoxicity

Cytotoxicity was determined using L929 (mouse fibroblast, subcutaneous connective tissue) following the method of Benjakul, Karnjanapratum, and Visessanguan [21] with slight modifications. Briefly, normal L929 cell lines were cultivated in DMEM and supplemented with 10% fetal bovine serum. The cells were then placed in a 5% CO_2 incubator at 37 °C, and cell line cultures with a density of 2×10^4 cells/mL were placed in microplate wells (100 μL per well). Subsequently, 100 μL of each sample at a concentration of 1 mg/mL was introduced into the wells and incubated (37 °C, 72 h). The cell viability assessment was performed using the MTT cell proliferation kit I (Roche Diagnostics; Burgess Hill, West Sussex, UK) and following the manufacturer's recommendations. Briefly, after the incubation period, 10 μL of the MTT labeling reagent (final concentration 0.5 mg/mL) were added to each well and incubated (4 h, 37 °C, 5% CO_2). Subsequently, 100 μL of solubilization buffer were added to each well, and the plates were incubated overnight (37 °C, 5% CO_2). The solubilization of the purple formazan crystals was measured by reading the absorbance of the wells at 570 nm with a spectrophotometer (BioTek, Winooski, VT, USA).

2.8. Antioxidant Activity

2.8.1. DPPH Radical Scavenging Activity

DPPH radical scavenging activity was evaluated using the method described by Ambigaipalan and Shahidi (2015) [22] with slight modifications. Briefly, 200 μL samples (1 mg/mL) were mixed with 800 μL of 0.1 mM DPPH solution in 95% methanol. The mixtures were incubated in dark conditions (30 min) and the absorbance of the reactions was read at 517 nm using a spectrophotometer (BioTek, Winooski, VT, USA). Distilled water and ascorbic acid were used as negative and positive controls, respectively.

The DPPH radical scavenging activity (%) of the samples was calculated using Equation (4).

$$\text{DPPH radical scavenging } (\%) = \frac{\text{AC} - \text{AS}}{\text{AC}} \times 100 \quad (4)$$

where A_C and A_S are the absorbance of the control and samples, respectively.

2.8.2. Ferrous Chelating (FC) Activity

FC activity was determined as detailed by T. Wang, Jonsdottir, and Ólafsdóttir (2009) [23]. Briefly, 1.74 mL of deionized water was added to 200 μL samples (1 mg/mL), the positive control (Na_3EDTA solution at 1 mg/mL), or blanks (water) and mixed with 20 μL of 2 mM $FeCl_2$ and 40 μL of 5 mM ferrozine solutions. The mixtures were shaken thoroughly for 10 min at room temperature. The absorbance of the reactions was measured at 562 nm using a spectrophotometer (BioTek, Winooski, VT, USA). Na_3EDTA served as the reference standard, and 100 μL of distilled water was employed instead of the samples as a control. Instead of the ferrozine solution, 10 μL of distilled water was utilized as a blank. The FC activity (%) of the samples was calculated using the following equation (Equation (5)).

$$\text{FC (\%)} = [A_C - (A_S - A_B)] \times 100 \quad (5)$$

where A_C refers to the absorbance of the control, A_S to the absorbance of the sample or standard, and A_B to the absorbance of the blank.

2.8.3. Hydroxyl Radical Scavenging Activity

Hydroxyl radical scavenging activity was determined following the method as described by B. Wang et al. (2013) [24]. Briefly, 2 mL samples (1 mg/mL) were mixed with 1.865 mM of a 1,10-phenanthroline solution at a ratio of 1:2 (v/v). Thereafter, 750 μL of $FeSO_4 \cdot 7H_2O$ solution and 1 mL of H_2O_2 (0.03% v/v) were added to the mixtures and incubated (37 °C, 1 h). The absorbance of the samples was recorded at 536 nm using a spectrophotometer (BioTek, Winooski, VT, USA). The negative control (n) and blank (b) were established using the reaction mixture without the inclusion of samples (s) and H_2O_2, respectively. The hydroxyl radical scavenging activity (%) of the samples was calculated using the following equation (Equation (6)).

$$\text{Hydroxyl radical scavenging (\%)} = \frac{As - An}{Ab - An} \times 100 \quad (6)$$

where A_s, A_n, and A_b refer to the absorbance of the sample, negative control, and blank, respectively.

2.8.4. Ferric Reducing Antioxidant Power (FRAP)

FRAP was determined as described by Oyaizu (1986) [25] with the modifications performed by Karawita et al. (2005) [26]. Briefly, 2.5 mL of each sample (1 mg/mL) or trolox (500 μM), as a positive control, was mixed with equal volumes of 200 mM phosphate buffer (pH 6.6) and 1% potassium ferricyanide. The mixtures were incubated (50 °C, 20 min) and allowed to cool down to room temperature; the addition of 2.5 mL of trichloroacetic acid (10%) followed. The mixtures were then centrifuged (4200× g, 10 min). An aliquot (5 mL) of the resultant supernatant was spiked with an equal volume of deionized water and 1 mL of ferric chloride solution (0.1%, w/v). The absorbance of the reactions was measured at 690 nm using a spectrophotometer (BioTek, Winooski, VT, USA).

2.9. *Functional Properties*

2.9.1. Water Holding and Fat Binding Capacities

The water holding capacity (WHC) and fat-binding capacity (FBC) were measured as described in Mirzapour-Kouhdasht, Moosavi-Nasab, Kim, and Eun (2021) [27]. For the WHC, 100 mg samples were combined with 10 mL of distilled water and the mixtures were gently agitated (100 rpm, 1 h, room temperature). Subsequently, these mixtures were centrifuged (2000× g, 25 min), the supernatants were drained, and the WHC was quantified by comparing the initial weight of the tubes to their weight after draining. In the case of the FBC, 100 mg samples were combined with 2 mL of olive oil and the mixtures were allowed to stand for 30 min at room temperature with intermittent shaking every 10 min. Subsequently, the mixtures were centrifuged (2000× *g*, 25 min) to drain the oil and the FBC was determined by comparing the initial weight of the tubes to their weight after draining.

2.9.2. Emulsion Activity and Stability

The emulsion activity index (EAI) and emulsion stability index (ESI) were assessed as reported by Mirzapour-Kouhdasht et al. (2021) [27]. A total of 30 mL of each sample (0.1% *w*/*v*) was blended with 10 mL of soybean oil using a homogenizer (CAT Unidrive 1000, CAT Scientific, Paso Robles CA, USA). After homogenization at 0 and 10 min, 100 μL of the emulsion was diluted 100-fold using SDS solution (0.1% *w*/*v*). The absorbance of the samples was then measured at 500 nm using a microplate reader. The EAI (m^2/g) and ESI (min) were calculated using Equation (7) and Equation (8), respectively.

$$\text{EAI}\ (m^2/g) = \frac{2 \times 2.303 \times A}{0.25 \times Cp} \tag{7}$$

$$\text{ESI}\ (\text{min}) = \frac{A0 \times \Delta t}{\Delta A} \tag{8}$$

where A denotes the absorbance of the mixtures at 500 nm; Cp denotes the protein concentration, A_0 refers to the initial absorbance, and A_{10} refers to the absorbance at 10 min post-homogenization. Δt corresponds to a time interval of 10 min, while ΔA represents the disparity between A_0 and A_{10}.

2.9.3. Foam Expansion and Stability

Foam expansion (FE) and foam stability (FS) were determined using the method described by Shahidi, Han, and Synowiecki (1995) [28]. Following the homogenization of each sample (2% *w*/*v*, 1 min, room temperature), the samples were transferred to cylindrical vessels. Alterations in the sample volume were gauged after 30 and 60 min. FE (%) and FS (%) were calculated using Equation (9) and Equation (10), respectively.

$$\text{FE}\ (\%) = (V_A - V_B)/(V_B) \times 100 \tag{9}$$

$$\text{FS}\ (\%) = (V_{30/60} - V_B)/(V_B) \times 100 \tag{10}$$

where V_A is the total volume after homogenizing, V_B is the total volume before homogenizing, and $V_{30/60}$ is the total volume after 30 and 60 min.

2.10. Statistical Analyses

All analyses were performed in duplicate, and cytotoxicity was assayed in triplicate. The normality of the data sets was confirmed by Kolmogorov–Smirnov tests ($p > 0.05$) and the equal variances were tested using Levene's tests ($p < 0.05$) indicating that the null hypothesis of equal variances could not be accepted. One-way multivariate analysis of variance (one-way MANOVA) was used to check differences in the proximate composition of the seeds based on the germination period. Two-way multivariate analysis of variance (MANOVA) was used to analyze the influence of germination, MW purification, and interaction between both factors in digestibility, ACE-I inhibitory activity, cytotoxicity, antioxidant activities, and functional properties. Games–Howell post hoc tests were used in all cases to analyze in detail the differences within the groups. All data were analyzed using SPSS version 23 (IBM, North Castle, NY, USA). In all cases, the criterion for statistical significance was $p < 0.05$.

3. Results and Discussions

3.1. Proximate Analyses

The proximate composition (moisture, protein, fat, carbohydrate, fiber, and ash contents) of lentils from days 0 to 6 during their germination is summarized in Table 1. There were statistical differences in all parameters during germination ($p < 0.001$; F = 2.990). Overall, the fat and carbohydrate contents in the lentils decreased significantly over the germination period, reaching minimum levels of ≈ 1% of fat and 37% of carbohydrates in days 5–6. An opposite behavior was appreciated in the case of the moisture, protein, fiber, and ash contents, with a tendency to increase during the process of germination, reaching maximum levels of approximately 12–13% moisture, 25% protein, 21% fiber, and 3–4% ash contents around days 4–5 of germination.

Table 1. Summary of changes in proximate composition of lentil seeds during germination (0–6 days).

Germination Period (Days)	Moisture (%)	Protein (%)	Fat (%)	Carbohydrates (%)	Fiber (%)	Ash (%)
0	9.33 ± 0.00^{e}	22.46 ± 0.01^{e}	2.51 ± 0.01^{a}	43.20 ± 0.01^{a}	19.95 ± 0.01^{e}	2.41 ± 0.02^{c}
1	10.18 ± 0.01^{d}	22.59 ± 0.02^{d}	2.30 ± 0.01^{a}	42.16 ± 0.01^{b}	20.16 ± 0.00^{d}	2.44 ± 0.04^{bc}
2	10.67 ± 0.01^{c}	22.91 ± 0.01^{d}	2.26 ± 0.00^{a}	41.07 ± 0.00^{c}	20.33 ± 0.01^{c}	2.57 ± 0.02^{bc}
3	10.91 ± 0.00^{bc}	23.34 ± 0.03^{c}	1.94 ± 0.00^{b}	40.38 ± 0.02^{d}	20.68 ± 0.01^{b}	2.69 ± 0.01^{b}
4	11.90 ± 0.05^{b}	24.37 ± 0.02^{b}	1.18 ± 0.00^{c}	38.19 ± 0.08^{e}	21.31 ± 0.02^{a}	3.00 ± 0.01^{a}
5	12.25 ± 0.05^{ab}	24.56 ± 0.04^{ab}	1.12 ± 0.01^{c}	37.44 ± 0.07^{e}	21.44 ± 0.03^{a}	3.44 ± 0.03^{a}
6	12.73 ± 0.02^{a}	24.71 ± 0.02^{a}	1.08 ± 0.01^{c}	36.66 ± 0.00^{e}	21.52 ± 0.00^{a}	3.52 ± 0.00^{a}

Data are expressed as average ± standard deviation (n = 2). Different letters within each column indicate statistically significant differences ($p < 0.05$) for each parameter during 6 days of germination.

These results are similar to those of Fouad and Rehab [14], who reported an increase in moisture, protein, ash, and fiber and a decline in carbohydrate and fat content in lentils during their germination. Moreover, Ghumman, Kaur, and Singh [29] reported an increase in protein contents in lentils during 4 days of germination. However, contrary to the findings of this study, the authors reported a decrease in the levels of ash during germination. The changes appreciated in the protein of seeds during germination could be related to the production of enzymes needed during the process of germination, leading to increased amino acid production and the degradation of high-MW proteins into small peptides [29].

Lentils, like other pulses of this family, may contain variable amounts of ANCs, including tannins, lectins, phytic acid, and protease inhibitors that can potentially reduce their nutritional benefits. However, the protein extraction processes used in this study (alkaline extraction and isoelectric precipitation) have been previously reported as being effective strategies to alleviate the adverse effects of ANCs and to improve the nutritional value and digestibility of these proteins. This could possibly be attributed to the denaturation or damage of ANCs at high pH values [9,30]. Moreover, the final neutral pH of the lentil protein isolates provides a net negative charge on the proteins, which can lead to weak protein–phytic acid binding and the leaching out of phytic acid during the protein extraction process [31].

3.2. Structural Characteristics of Protein Isolates

The structural conformation of proteins affect the functionality and digestibility of these compounds [32]. Thus, changes in the secondary structural composition of proteins can alter their technological properties [33] which are critical for their future uses in the food industry. The changes in the chemical bonds and structural properties of protein isolates during the process of lentil germination were analyzed by FTIR spectroscopy (Figure 3).

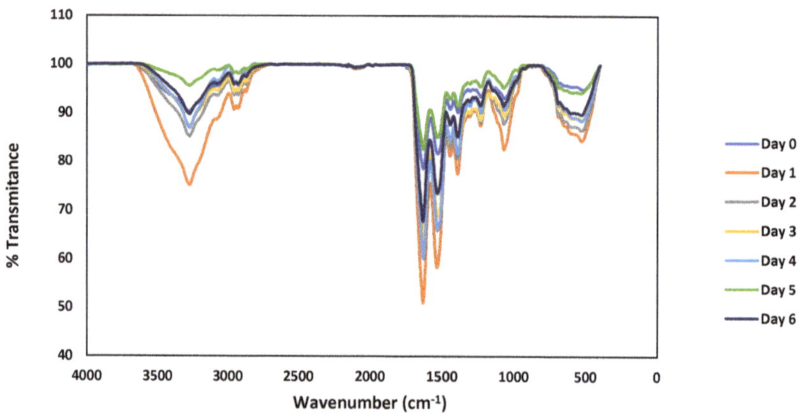

Figure 3. FTIR spectra of protein isolates generated from lentil seeds during germination (days 0–6). Results were recorded three times for each sample.

Previous studies analyzing the FTIR spectra of germinated pulses have reported enhancement in the water-solubility of proteins for enzymatic activity to develop new tissues during germination [34]. The vibration of peptide bonds at the wavelength of 1700–1500 cm^{-1} indicates the protein content of lentil seeds [34]. The amide I (C=O, O-H) region (1730–1580 cm^{-1}) and the amide II (N-H, C-N) region (1580–1470 cm^{-1}) were analyzed to determine the relative proportions of different secondary structures by computing areas of the spectrums at ≈1630 cm^{-1} (β-sheet) as the main constituents of the protein's secondary structure (see Table 2). Previous studies have reported that changes in the amide I region depict changes in the secondary structure of protein molecules [35].

Table 2. Content of β-sheet of lentil protein during 6 days of germination.

Germination Period (Days)	Area Under Curve (AUC)	% β-Sheet
0	8.39	15.78
1	21.53	15.79
2	16.76	17.05
3	14.56	16.94
4	16.08	19.18
5	6.40	18.45
6	12.48	18.17

The results of this study indicate a reduction in the β-sheet configuration of lentil proteins during germination, which is consistent with previous observations on the structural modifications of proteins during the germination of pulses. For example, the reduction in β-sheet content might correspond with increased protein solubility and digestibility, enhancing the bioavailability of nutrients. These structural changes can improve the functional properties of proteins, including changes in emulsification, foaming, and water-holding capacity, which can be of benefit for various food applications. However, it may also lead to reduced thermal stability and the potential loss of some functional properties that rely on the β-sheet structure, like gel formation [34,35]. Thus, the functional properties of these proteins were also analyzed, aiming to provide a complete range of future applications for these proteins in food formulation depending on the germination stage of the seeds.

Table 2 shows the changes in the content of β-sheets in lentil proteins during 6 days of germination. After an initial increase, there was a notable reduction in β-sheets that stabilized around days 4–6. This trend suggests a dynamic restructuring process where the initial breakdown of complex structures facilitates new tissue development and enzymatic activities, ultimately leading to a more flexible and functional protein configuration suitable for diverse food industry applications. In summary, the observed reduction in β-sheet configuration during lentil germination aligns with enhanced functional properties, making germinated lentil proteins potentially more versatile for food industry applications, while balancing improvements in digestibility and nutrient bioavailability.

3.3. *Nutritional Properties*

3.3.1. Digestibility

Aiming to elucidate the impact of germination on protein digestibility, the lentil protein isolates were purified into different MW fractions, and the digestibility of different MW fractions achieved from protein isolates from lentils during the process of germination (days 0–6) are compiled in Table 3. The digestibility was affected by germination ($p < 0.001$; F = 73683.102), MW purification ($p < 0.001$; F = 3709.894), and the interaction between both factors (germination*MW; $p < 0.001$; F = 67.020).

Table 3. Digestibility (%) of different MW fractions achieved from protein isolates generated from lentils during process of germination (days 0–6).

Germination Duration (Day)	MW (kDa)		
	MW > 5	3 < MW < 5	MW < 3
0	16.32 ± 0.01 ^{Bg}	17.00 ± 0.01 ^{Ag}	17.04 ± 0.00 ^{Ag}
1	17.05 ± 0.01 ^{Cf}	17.37 ± 0.00 ^{Bf}	18.26 ± 0.03 ^{Af}
2	18.75 ± 0.00 ^{Ce}	19.22 ± 0.02 ^{Be}	19.78 ± 0.01 ^{Ae}
3	20.67 ± 0.03 ^{Cd}	21.45 ± 0.02 ^{Bd}	21.85 ± 0.04 ^{Ad}
4	22.64 ± 0.02 ^{Bc}	22.89 ± 0.01 ^{Bc}	23.16 ± 0.01 ^{Ac}
5	23.23 ± 0.01 ^{Bb}	23.80 ± 0.00 ^{Ab}	24.16 ± 0.02 ^{Ab}
6	24.92 ± 0.02 ^{Ca}	25.64 ± 0.02 ^{Ba}	26.05 ± 0.03 ^{Aa}

Data are expressed as average ± standard deviation (n = 2). Different uppercase letters indicate statistically significant differences ($p < 0.05$) in digestibility between fractions of different MW within same germination day. Lowercase letters indicate statistically significant ($p < 0.05$) differences in digestibility between different germination days at same MW.

The protein digestibility after the dialysis of three MW fractions of lentil protein isolates ranged from 16.32 to 17.04%. As observed in Table 3, the digestibility of all tested fractions was increased by pepsin and pancreatin hydrolysis during sprouting. The digestibility of the fractions with MW > 5 kDa and 3 kDa < MW < 5 kDa and MW < 3 kDa of lentil seeds increased significantly ($p < 0.05$) from levels of 16.32, 17, and 17.04% to reach levels of 24.92, 25.64, and 26.05% after 6 days of germination, respectively. This indicates an increased susceptibility of the peptide bonds to digestive enzymes in the proteins as the germination of the lentil seeds progresses.

As previously mentioned, the interaction between MW and germination time was significant, as the rate of increase in digestibility over time varied across different MW fractions. The low-MW fractions had a more pronounced improvement in digestibility, indicating that both MW and germination duration significantly influence protein digestibility. The observed increase in protein digestibility during germination aligns with previous studies. Thereby, Di, Li, Chang, Gu, Duan, Liu, Liu, and Wang [35] reported similar findings and attributed these changes to enhanced protease activity, the deactivation of protease inhibitors, and protein degradation in small and soluble peptides [36]. This enhanced digestibility of lentil proteins can improve their nutritional quality and functionality for food applications, potentially making protein ingredients extracted from germinated lentils more desirable for health-conscious consumers and food manufacturers seeking ingredients with high bioavailability and improved functional properties [37].

3.3.2. ACE-I Inhibitory Activity

The inhibition of ACE-I (EC 3.4.15.1) is an effective strategy for lowering blood pressure. Multiple peptides have been identified from various natural sources as inhibitors of ACE-I with potential to replace conventional drugs, such as captopril, and avoid the adverse side effects of pharmacological treatments [38].

The ACE-I inhibitory activity of different MW fractions from protein isolates from lentils during the process of germination (days 0–6) are summarized in Table 4. There were statistical differences in ACE-I inhibitory activity during germination ($p < 0.05$; F = 491.709), MW purification ($p < 0.05$; F = 1344.830), and the interaction between both factors (germination*MW; $p < 0.05$; F = 10.172). Overall, the IC_{50} of the compounds had its lowest levels in fractions of low MW within the same period of germination. Moreover, independently of the MW of the fractions analyzed, the IC_{50} of the compounds decreased with an increased germination time. The lowest activity was found in non-germinated lentil protein (IC_{50}~1.49–1.03%), and the IC_{50} of the compounds decreased to reach 0.95-0.69% after a week of germination. Thus, a low MW and germination time had a positive effect, increasing the ACE-I inhibitory activity of lentil proteins. Bamdad et al. [39] also reported that germination can release bioactive peptides with high ACE inhibitory activity. The increase in the ACE-I inhibitory activity of germinated seeds can be explained by the increased degradation of protein structures, also appreciated in the previously described FTIR results. Furthermore, previous studies have also reported an increased activity of endogenous proteases during sprouting, leading to the generation of peptides with specific MW, composition, and amino acid sequences with potential ACE-I inhibitory characteristics [40].

Table 4. ACE-I inhibitory activity (IC_{50}) of different MW fractions achieved from protein isolates from lentils during the process of germination (days 0–6).

Germination Period (Days)	MW (kDa)		
	MW > 5	3 < MW < 5	MW < 3
0	1.49 ± 0.02^{Aa}	1.19 ± 0.01^{Ba}	1.03 ± 0.00^{Ba}
1	1.39 ± 0.00^{Aa}	1.17 ± 0.02^{Aa}	0.97 ± 0.01^{Bab}
2	1.28 ± 0.01^{Aab}	1.11 ± 0.01^{Ba}	0.91 ± 0.00^{Cab}
3	1.21 ± 0.01^{Ab}	1.05 ± 0.01^{Ba}	0.87 ± 0.01^{Cb}
4	1.05 ± 0.01^{Ac}	0.91 ± 0.00^{Bb}	0.73 ± 0.02^{Bc}
5	1.00 ± 0.01^{Ac}	0.88 ± 0.01^{Bbc}	0.71 ± 0.01^{Cc}
6	0.95 ± 0.01^{Ac}	0.83 ± 0.00^{Ac}	0.69 ± 0.01^{Bc}

Data are expressed as average ± standard deviation (n = 2). Different uppercase letters indicate statistically significant differences ($p < 0.05$) in ACE-I inhibitory activity between fractions of different MW within same germination day. Lowercase letters indicate statistically significant ($p < 0.05$) differences in ACE-I inhibitory activity between different germination days at same MW.

The IC_{50} levels of the fractions generated in this study were similar to those reported in previous research dealing with the hydrolysis of protein isolates from lentils and other pulses. In such research, protein hydrolysates generated from lentil protein hydrolysates had an IC_{50} of 0.25 mg/mL [41], red lentil protein hydrolysates had an IC_{50} of 0.44 mg/mL [42], green soybean protein hydrolysates had IC_{50} of 0.14 mg/mL [43], and lupin protein hydrolysates had an IC_{50} of 0.22 mg/mL [44].

3.3.3. Cytotoxicity

The effect of protein isolates of different MW fractions achieved from protein isolates from lentils in different germination stages on the proliferation of L929 cells is represented in Figure 4. The cytotoxicity of the samples was affected by the germination ($p < 0.001$; F = 192.555), MW purification ($p < 0.001$; F = 2960.312), and the interaction between both factors (germination*MW; $p < 0.001$; F = 101.532). Protein fractions of MW < 3 kDa and 3 kDa < MW < 5 kDa showed significant adverse effects. The cell proliferation remained unchanged in the presence of high-MW protein fractions (MW > 5 kDa) during the sprouting process. The lowest cell proliferation of the L929 cell line (97.63%) was related to the treatment with the MW < 3 kDa lentil protein fraction after 6 days of germination. Meaningful differences were appreciated among cell lines treated with different MW fractions achieved during germination and thus the germination process had a toxicity effect of MW < 3 kDa and 3 kDa < MW < 5 kDa fractions on normal L929 cells.

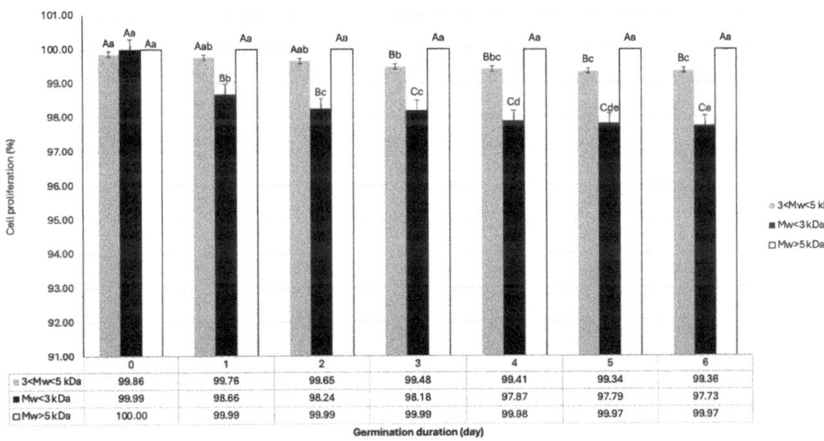

Figure 4. Cytotoxicity of different MW fractions achieved from protein isolates from lentils during process of germination (days 0–6) using L929 cell lines. Different uppercase letters indicate statistically significant differences ($p < 0.05$) in cytotoxicity between fractions of different MW within same germination day. Lowercase letters indicate statistically significant ($p < 0.05$) differences in cytotoxicity between different germination days at same MW. Data are shown as average of three replicates.

In a study by Hlosrichok and Aunpad [45], winged bean (*Psophocarpus tetragonolobus*) seeds' protein was extracted by gamma rays and this was followed by enzymatic hydrolysis with alcalase (55 °C, 6 h). The study revealed that winged bean protein hydrolysates had no toxicity against L929 cells. These variable results could be attributed to the different protein sources used in each study. There are some types of proteins and glycoproteins in lentils called lectins that can induce cytotoxicity in L929 cell lines [46]. To the best of our knowledge, there is not enough evidence of using L929 cell lines to examine the effect of germination on the cytotoxicity of proteins derived from lentil seeds. Hence, further studies are needed to clarify the exact effects of this process and the compounds generated on cell proliferation.

3.4. *Antioxidant Activities*

The antioxidant activities of three fractions of different MW achieved from protein isolates from lentils during the first 6 days of the germination period were assessed in terms of their ability to scavenge DPPH radicals (Figure 5a), ferrous chelating activity (Figure 5b), hydroxyl radical scavenging activity (Figure 5c), and FRAP (Figure 5d). There were statistical differences in all antioxidant activities tested on the basis of germination ($p < 0.001$; F = 54.784), MW ($p < 0.001$; F = 12846.455), and the interaction between both (germination*MW; $p < 0.001$; F = 9.975).

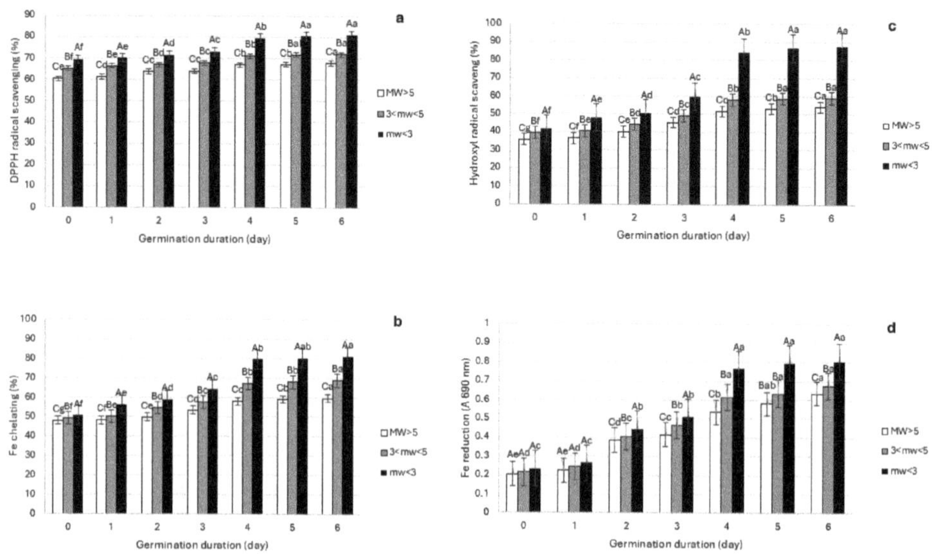

Figure 5. Antioxidant activities: (**a**) DPPH (%), (**b**) Fe chelating (%), (**c**) hydroxyl radical (%), and (**d**) FRAP (A) activities of different MW fractions achieved from protein isolates from lentils during process of germination (days 0–6). Different uppercase letters indicate statistically significant differences ($p < 0.05$) in antioxidant activity between fractions of different MW within same germination day. Lowercase letters indicate statistically significant ($p < 0.05$) differences in antioxidant activity between different germination days at same MW. Data are shown as average of two replicates.

Although low-MW fractions already had high DPPH scavenging activity before the process of germination started (<70%), this activity increased to reach over 80% after 6 days of germination (Figure 5a). This increment in antioxidant activity was not appreciated during the process of germination in the case of fractions of high MW (60% and 67% before and after germination, respectively). However, the increment was statistically significant ($p < 0.05$) when comparing the DPPH radical scavenging activity of the lower-MW fractions of the protein to non-germinated and to germinated ones. These results are aligned with those described by Fouad and Rehab [14], showing a 53% increase in radical scavenging activity after a week of the germination of lentil seeds. Moreover, when studying the antioxidant ability of different seeds during germination, previous reports have emphasized an increased level of DPPH scavenging activity during a germination period of 4–7 days in soybean, dill, and anise seeds [47]. The DPPH radical scavenging activity of various peptides achieved using protein hydrolysis have been previously reported to depend on the MW size and sequence of the peptides. Both factors are also influenced by the type of enzyme used to generate those hydrolysates [48] or the endogenous hydrolytic activities present naturally in the seeds, as well as the presence of other natural antioxidants, such as vitamin C and tocopherols [14]. Previous studies with pulses have demonstrated high levels of DPPH in protein fractions of low MW. Olagunju, Omoba, Enujiugha, Alashi, and Aluko [48] reported the highest DPPH radical scavenging activity in low-MW pancreatin–hydrolysate peptides achieved from pigeon pea protein isolates.

In the case of the ferrous chelating power of the compounds (Figure 5b), this activity can also be affected by the peptide size and amino acid composition [27]. The results of the ferrous chelating activity of lentil seeds demonstrated a marked increase in the Fe^{2+} chelating ability on day 2 of germination for all MW fractions. This increase was significantly greater in low-MW protein fractions (<3 kDa), reaching 58.97%, in comparison with the other two higher-MW fractions (54.51% in 3 < MW < 5 kDa and 49.74% in MW > 5 kDa). Balanescu, Busuioc, Botezatu, Gosav, Avramescu, Furdui, and Dinica [47] studied the reducing iron chelating power of soybean, dill, and anise seeds and their sprouts (germinated for 4 to 7 days). The authors reported that while the extracts from sprouts had a higher iron binding ability than the seeds, this activity was also influenced by the type of biomass studied. Soybean had the highest reducing iron chelating power, followed by dill and anise seeds, a fact that could be explained by their differences in their amino acid composition [47].

The hydroxyl radical scavenging activity followed a similar trend to that previously mentioned in ferrous chelating activity (Figure 5c). From day 2 of germination, all protein fractions showed a statistical increase in their capacity to scavenge hydroxyl radicals, being this increased antioxidant properties more accentuated in low-MW fractions (MW < 3 kDa). In raw lentil seeds, the hydroxyl radical scavenging was 35, 39, and 41% for fractions with MW > kDa 5 and 3 kDa < MW < 5 kDa and MW < 3 kDa; that increased to 54, 59, and 87% after 6 days of germination, respectively. The reaction between H_2O_2 and Fe^{2+} produces transitory and intensely reactive hydroxyl radicals with oxidative damage effects on biological macromolecules, such as DNA, proteins, and lipids [48]. Thus, antioxidants with high hydroxyl radical scavenging ability can significantly prevent oxidative stress reactions in living cells [48]. Previous studies have also reported an influence of the amino acid composition of peptides on their hydroxyl radical scavenging ability. Thereby, a higher amount of hydrophobic amino acids have been related to the higher hydroxyl radical scavenging ability of these peptides [27]. Moreover, high hydroxyl scavenging activity was reported in low-MW peptides. Thereby, the highest hydroxyl radical scavenging ability of pigeon pea protein hydrolysates was appreciated in peptides of low MW (<1 kDa) generated from ultrafiltration from full hydrolysates [48].

Measuring FRAP is another indicator of the antioxidant ability of different compounds. The presence of hydrogen or electron donor compounds (reductones) can reduce lipid peroxyl radicals and inhibit lipid oxidation [14,48]. The reported positive correlation between FRAP, hydroxyl radical scavenging, ferrous chelating activity, and the antioxidant activity of different peptide fractions was consistent with previous findings [49]. In a study conducted by Parit et al. [50], a proteomic analysis was performed on wheat (*Triticum aestivum*) seeds at 0, 8, and 16 days of germination. The study revealed that the seeds germinated for 16 days exhibited the highest FRAP value, followed by those germinated for 8 days and then the ungerminated seeds (0 days). This indicates a direct correlation between the duration of germination and the FRAP value, with longer germination periods leading to higher antioxidant capacity. As shown in Figure 5d, the FRAP of lentil protein fractions increased during the process of germination. The absorbance of low-MW fractions increased from 0.23 to 0.8 after 6 days of germination. Investigating the antioxidant activity of *Amaranthus viridis* seeds, the DPPH, FRAP, and metal chelating ability showed a remarkable rise during the process of the germination of the seeds during a period of 3 days [51]. Moreover, Fouad and Rehab [14] also reported an increase in the FRAP from 0.22 in raw lentil seeds to 0.55 in lentil seeds germinated for 6 days. In contrast to the results of this work, Olagunju, Omoba, Enujiugha, Alashi, and Aluko [48] reported a higher FRAP in high-MW peptides (>10 kDa) than in low-MW ones (1–10 kDa) in pigeon peas, attributed to differences in the amino acid composition of the different bioactive peptide fractions generated during the germination process [27].

3.5. Functional Properties

Lentil protein ingredients are generally described as having acceptable functional properties, allowing them to retain water and fat in food formulations and to be used as a promising emulsifier and foaming agents. However, these properties are influenced by changes in MW and surface hydrophobicity that the compounds experience during the process of protein hydrolysis [52]. A summary of the functional properties of different MW fractions achieved from protein isolates from lentils during the process of germination (days 0–6) in this study is provided in Table 5. There were statistical differences in all functional properties tested on the basis of germination ($p < 0.001$; F = 25.004), MW ($p < 0.001$; F = 7400.944), and the interaction between both (germination*MW; $p < 0.001$; F = 207.280).

Table 5. Functional properties of different MW fractions generated from protein isolates from lentils during process of germination (days 0–6).

Germination Period (Days)	MW (kDa)	WHC (%)	FBC (%)	EAI (m²/g)	ESI (min)	FE (%)	FS (%)	
							30 min	60 min
0	MW > 5	93.24 ± 0.02^{Fc}	97.67 ± 0.02^{Gc}	101.32 ± 0.02^{Gc}	25.32 ± 0.02^{Aa}	124.01 ± 0.07^{Ga}	67.10 ± 0.01^{Ga}	54.60 ± 0.01^{Ga}
0	3 < MW < 5	95.38 ± 0.01^{Eb}	99.35 ± 0.01^{Gb}	103.66 ± 0.03^{Gb}	24.17 ± 0.01^{Ab}	123.18 ± 0.01^{Da}	64.34 ± 0.01^{Gb}	52.11 ± 0.01^{Gb}
0	MW < 3	97.89 ± 0.01^{Ga}	106.31 ± 0.01^{Ga}	105.12 ± 0.01^{Fa}	22.18 ± 0.03^{Ac}	121.35 ± 0.01^{Gb}	63.17 ± 0.01^{Gc}	51.45 ± 0.01^{Gc}
1	MW > 5	94.68 ± 0.00^{Ec}	98.65 ± 0.01^{Fc}	102.32 ± 0.02^{Fb}	25.02 ± 0.04^{ABa}	126.84 ± 0.04^{Fa}	69.47 ± 0.02^{Fa}	55.12 ± 0.01^{Fa}
1	3 < MW < 5	95.16 ± 0.01^{Eb}	103.24 ± 0.01^{Fb}	106.76 ± 0.35^{Fa}	24.23 ± 0.01^{Ab}	124.05 ± 0.01^{Db}	66.22 ± 0.02^{Fb}	53.28 ± 0.01^{Fb}
1	MW < 3	98.37 ± 0.02^{Fa}	108.96 ± 0.02^{Fa}	109.55 ± 0.01^{EFa}	22.08 ± 0.01^{Ac}	122.73 ± 0.02^{Fc}	65.81 ± 0.01^{Fc}	52.25 ± 0.04^{Fc}
2	MW > 5	95.09 ± 0.01^{Dc}	103.24 ± 0.02^{Ec}	105.44 ± 0.01^{Ec}	24.98 ± 0.01^{Ba}	128.13 ± 0.01^{Ea}	71.18 ± 0.02^{Ea}	59.49 ± 0.02^{Ea}
2	3 < MW < 5	96.36 ± 0.03^{Db}	108.46 ± 0.03^{Eb}	108.66 ± 0.02^{Eb}	23.44 ± 0.02^{Bb}	125.18 ± 0.00^{Db}	68.58 ± 0.04^{Eb}	55.16 ± 0.03^{Eb}
2	MW < 3	97.05 ± 0.01^{Ea}	111.72 ± 0.03^{Ea}	112.09 ± 0.01^{Ea}	22.04 ± 0.02^{Ac}	122.97 ± 0.02^{Ec}	66.01 ± 0.01^{Ec}	53.01 ± 0.03^{Ec}
3	MW > 5	97.12 ± 0.02^{Dc}	108.36 ± 0.01^{Dc}	109.05 ± 0.01^{Dc}	24.70 ± 0.01^{Ca}	130.86 ± 0.03^{Da}	75.83 ± 0.04^{Da}	64.32 ± 0.01^{Da}
3	3 < MW < 5	97.86 ± 0.01^{Cb}	113.45 ± 0.01^{Db}	113.25 ± 0.01^{Db}	23.10 ± 0.01^{Cb}	127.65 ± 0.01^{Cb}	70.21 ± 0.01^{Db}	58.50 ± 0.03^{Db}
3	MW < 3	98.29 ± 0.01^{Da}	117.24 ± 0.01^{Da}	118.46 ± 0.02^{Da}	21.74 ± 0.01^{Bc}	124.11 ± 0.01^{Dc}	69.45 ± 0.02^{Dc}	54.13 ± 0.01^{Dc}
4	MW > 5	97.51 ± 0.02^{Cc}	109.33 ± 0.02^{Cc}	114.80 ± 0.01^{Cc}	24.16 ± 0.02^{Da}	133.44 ± 0.05^{Ca}	84.53 ± 0.02^{Ca}	73.14 ± 0.01^{Ca}
4	3 < MW < 5	98.47 ± 0.01^{Bb}	116.65 ± 0.03^{Cb}	118.07 ± 0.01^{Cb}	22.91 ± 0.02^{Cb}	130.12 ± 0.03^{Bb}	76.92 ± 0.01^{Cb}	68.37 ± 0.01^{Cb}
4	MW < 3	99.815 ± 0.03^{Ca}	121.33 ± 0.02^{Ca}	124.96 ± 0.01^{Ca}	20.53 ± 0.02^{Cc}	128.03 ± 0.02^{Cc}	72.14 ± 0.01^{Cc}	60.19 ± 0.01^{Cc}
5	MW > 5	98.22 ± 0.01^{Bc}	116.29 ± 0.01^{Bc}	119.35 ± 0.01^{Bc}	23.95 ± 0.02^{Da}	135.07 ± 0.01^{Ba}	87.27 ± 0.03^{Ba}	75.08 ± 0.01^{Ba}
5	3 < MW < 5	101.35 ± 0.03^{Ab}	120.78 ± 0.01^{Bb}	124.55 ± 0.02^{Bb}	22.45 ± 0.01^{Db}	134.77 ± 0.02^{Ab}	80.36 ± 0.02^{Bb}	71.08 ± 0.01^{Bb}
5	MW < 3	108.56 ± 0.01^{Ba}	127.35 ± 0.02^{Ba}	131.03 ± 0.04^{Ba}	20.23 ± 0.02^{CDc}	133.42 ± 0.02^{Bc}	74.51 ± 0.01^{Bc}	68.42 ± 0.01^{Bc}
6	MW > 5	100.36 ± 0.35^{Ac}	119.44 ± 0.01^{Ac}	123.63 ± 0.02^{Ac}	23.72 ± 0.01^{Ea}	137.97 ± 0.05^{Aa}	89.34 ± 0.03^{Aa}	79.61 ± 0.02^{Aa}
6	3 < MW < 5	109.55 ± 0.10^{Ab}	125.28 ± 0.03^{Ab}	129.13 ± 0.01^{Ab}	22.29 ± 0.02^{Db}	135.12 ± 0.02^{Ab}	82.43 ± 0.05^{Ab}	76.95 ± 0.02^{Ab}
6	MW < 3	117.64 ± 0.10^{Aa}	134.87 ± 0.03^{Aa}	135.42 ± 0.01^{Aa}	20.14 ± 0.02^{Dc}	133.87 ± 0.02^{Ac}	79.34 ± 0.05^{Ac}	72.13 ± 0.02^{Ac}

Data are expressed as average ± standard deviation (n = 2). Different uppercase letters indicate statistically significant ($p < 0.05$) differences between various germination days at same molecular weight. Lowercase letters indicate statistically significant differences ($p < 0.05$) in functional properties between fractions of different molecular weight within same germination day. Abbreviations in table are as follows: molecular weight (MW), water holding capacity (WHC), fat binding capacity (FBC), emulsion activity index (EAI), emulsion stability index (ESI), foam expansion (FE), and foam stability (FS).

3.5.1. WHC and FBC

WHC and FBC are influenced by the polarity of the amino acids present in proteins/peptides [53]. WHC refers to proteins' capacity to inhibit the liquid leakage of a product while being produced and/or stored [2]. The results of this study revealed that low-MW fractions had a higher ability to retain water, and there were significant differences between the same fractions during the process of germination until day 6, in which the low-MW fractions reached a maximum WHC level of ≈118%. Similarly, the FBC of protein isolates increased during germination and, after 6 days, the FBC of the protein isolates increased by more than 20%. Moreover, meaningful differences were observed in the ability of protein fractions with MW > 5 kDa and 3 kDa < MW < 5 kDa in holding oil, which could be due to the approximate unequal proportion of polar amino acids. Mirzapour-Kouhdasht, Moosavi-Nasab, Kim, and Eun [27] reported that the dual character of MW < 3 kDa fractions could be an asset in food systems to act as emulsifiers. Ghumman, Kaur, and Singh [29] reported similar results to those of this study. The authors outlined a significant increase in the WHC of germinated lentils' and horsegrams' flours that could be attributed to a rise in the content of low-MW proteins and polar groups upon germination. Other studies have also reported slight improvements in WHC in yellow pea and faba bean flours after soaking and germination (3 days) due to the exposure of hydrophilic moieties; however, few changes were observed in the case of their FBC [36].

3.5.2. EAI and ESI

The EAI and ESI of proteins are important functional properties when these ingredients are needed to act as surfactants in emulsions for their use in different food formulations. The results of this study indicate that EAI values increased during germination, while the MW was inversely correlated to the EAI values. Fractions of MW < 3 kDa had the highest EAI levels of 135.42 m^2/g after 6 days of germination, compared to any other fraction of high MW and any other germination period. However, the stability of the emulsions formed was also at its lowest (20 min). High-MW fractions were able to achieve high ESI levels, with germination having no effect on the stability of the emulsions until day 2 of germination. Setia, Dai, Nickerson, Sopiwnyk, Malcolmson, and Ai [36] analyzed the impact of germination in pulse's flours, reporting an improved EAI and ESI for the flours to a certain degree, due to the partial unfolding and dissociation of proteins, promoting surface activity. This effect could be attributed to an increased surface hydrophobicity of the compounds, which plays a key role in boosting emulsifying activities. Regarding the influence of the MW of the proteins, previous studies have reported that low-MW peptides can disperse oil globules more evenly by unfolding and rearranging the molecules at the oil–water interface, resulting in more surface hydrophobicity and a higher EAI [27]. Contrary to the results of this study, Ghumman, Kaur, and Singh [29] reported a decrease in the EAI with an increase in germination duration in lentil seeds, which could be attributed to an increase in β-sheets and a decrease in α-helix structures in the proteins. The inability of low-MW peptides to create a strong film around oil droplets at the oil–water boundary could lead to the reduced stability of the emulsion being produced [27].

3.5.3. FE and FS

Foams are created by the adsorption of proteins in the air–water interface, resulting in a decrease in the surface tension. During the foaming process, proteins need to unfold and create a supportive film around air bubbles, helping to maintain the foam's structure and prevent it from collapsing [2]. The FE of different MW fractions of ungerminated lentils ranged from 121.35 to 124.01% and these levels increased to 133.87–137.97% after 6 days of germination. In general, high-MW fractions of lentil protein isolates had higher FE ability compared to low-MW fractions, although the FE of these later fractions significantly increased ($p < 0.05$; F = 121049.217) during the process of germination. This fact could be explained by the higher capacity of high-MW peptides to interact at the air–water interface, in comparison to low-MW fractions [27]. Liu et al. [54] reported that germinating hemp seeds enhanced their foaming capacity. Specifically, the foaming capacity increased from levels of 69% recorded after 1 day of germination (comparable to non-germinated seeds) to levels of 111 and 107%, appreciated after 3 and 5 days of germination, respectively. This improvement could be attributed to increased protein solubility leading to an increased stabilization of air bubbles by the proteins.

In the case of FS values, the ability of foams to retain air in the form of bubbles for 30/60 min increased during germination. The results demonstrated that around 80–90% of the created foam was stable after 30 min and 70–80% of bubbles remained unchanged even after 60 min. Moreover, the fractions of low MW showed overall the lowest levels of FS compared to the high-MW counterparts, although the FS of these low-MW compounds also increased during the germination process. There are multiple studies reporting increased FS by limiting the hydrolysis of compounds [52,55,56]. Ghumman, Kaur, and Singh [29] reported that the foam capacity of lentil flour improved progressively due to germination (4 days), which could be attributed to the formation of β-sheets and the rearrangement of the α-helix in the proteins. To a large extent, the FS is affected by the surface rheological properties of interfacial films [52]. The observed increase in the FS of protein isolates of germinated lentils may be attributed to the partial unfolding of protein structures that occurred due to hydrolytic enzymes during sprouting. Enzymes activated by germination cause protein size reduction and the greater exposure of hidden hydrophobic groups to form stable films at the air–water interface [57]. Martínez, Sánchez, Patino, and Pilosof [52] reported that low degrees of hydrolysis (2%) in soy protein isolates enhanced surface activity and decreased the phase angle of the films which led to a delay in the collapse of the foam. However, further hydrolysis resulted in lower foam stability. Thus, it is likely that the limited hydrolysis of lentil proteins during the process of germination is probably responsible for the increased hydrophobicity of the compounds, allowing the formation of a stable surface film and enhancement of foam stability.

4. Conclusions

The current research into the ACE-I inhibitory activity, antioxidant activity, protein digestibility, cytotoxicity, and functional properties of distinct MW fractions of proteins isolated from lentil seeds over a 6-day period of germination yielded valuable insights into the functionality and future uses of these compounds. Overall, the protein digestibility, ACE-I inhibitory activity, and antioxidant activities (DPPH, ferrous chelating, hydroxyl radical, and FRAP activities) improved during the process of germination, with all of them being at their highest levels in the low-MW fractions (<3 kDa). Moreover, the cytotoxicity analysis using L929 cell lines in this study demonstrated that the germinated lentil proteins at low MWs exhibited a significant adverse effect on cell proliferation, whereas in the MW > 5 kDa fraction, no toxicity was observed. When exploring the functional properties of the lentil protein fractions, alterations in the WHC and FBC were appreciated during the process of germination that could be attributed to changes in the MW of the compounds during the process of germination. The emulsifying (EAI and ESI) and foaming abilities (FE and FS) were interesting as low-MW fractions had higher EAIs, while at the same time, they had considerably lower ESIs when compared to high-MW fractions; meanwhile, the FE and FS were both higher in high-MW fractions as compared to the low-MW ones. Nevertheless, further studies are necessary using the INFOGEST method to facilitate the comparison of the results. Further studies are also necessary to determine the specific mechanisms responsible for these changes and to analyze all aspects of their potential effects on human health and their interactions with other ingredients in other complex food formulations.

Author Contributions: Conceptualization, A.M.-K.; methodology, A.M.-K., S.S., M.M., J.-Y.H. and M.G.-V.; software, A.M.-K. and M.G.-V.; validation, A.M.-K., M.M., J.-Y.H. and M.G.-V.; formal analysis, A.M.-K.; investigation, A.M.-K.; resources, M.G.-V. and M.M.; data curation, M.G.-V. and A.M.-K.; writing—original draft preparation, A.M.-K. and S.S.; writing—review and editing, A.M.-K., S.S., M.M., J.-Y.H. and M.G.-V.; visualization, A.M.-K., S.S., M.M., J.-Y.H. and M.G.-V.; supervision, A.M.-K., M.M. and M.G.-V.; project administration, A.M.-K., M.M. and M.G.-V.; funding acquisition, M.G.-V. All authors have read and agreed to the published version of the manuscript.

Funding: The authors acknowledge the funding received from AMBROSIA funded by the Department of Agriculture Food and the Marine (DAFM) under the umbrella of the European Joint Programming Initiative "A Healthy Diet for a Healthy Life" (JPI-HDHL) and of the ERA-NET Cofund ERA HDHL (GA No 696295 of the EU Horizon 2020 Research and Innovation Program).

Institutional Review Board Statement: Not applicable.

Informed Consent Statement: Not applicable.

Data Availability Statement: All data generated or analyzed during this study are included in this published article.

Conflicts of Interest: The authors declare no conflicts of interest.

References

1. Shrestha, S.; van't Hag, L.; Haritos, V.; Dhital, S. Comparative study on molecular and higher-order structures of legume seed protein isolates: Lentil, mungbean and yellow pea. *Food Chem.* **2023**, *411*, 135464. [CrossRef] [PubMed]
2. Jarpa-Parra, M. Lentil protein: A review of functional properties and food application. An overview of lentil protein functionality. *Int. J. Food Sci. Technol.* **2018**, *53*, 892–903. [CrossRef]
3. Joshi, M.; Adhikari, B.; Aldred, P.; Panozzo, J.; Kasapis, S. Physicochemical and functional properties of lentil protein isolates prepared by different drying methods. *Food Chem.* **2011**, *129*, 1513–1522. [CrossRef]
4. S. Santos, C.; Silva, B.; MP Valente, L.; Gruber, S.; W. Vasconcelos, M. The effect of sprouting in lentil (*Lens culinaris*) nutritional and microbiological profile. *Foods* **2020**, *9*, 400. [CrossRef] [PubMed]
5. Khamidova, F.; Yormatova, D.; Majidov, K.K.; Bokijonov, M. Features of physico-chemical parameters of local varieties of lentil grains in Uzbekistan. *IOP Conf. Ser. Earth Environ. Sci.* **2022**, *1068*, 012023. [CrossRef]
6. Samaranayaka, A. Lentil: Revival of poor man's meat. In *Sustainable Protein Sources*; Elsevier: Amsterdam, The Netherlands, 2017; pp. 185–196.
7. Lopez, M.J.; Mohiuddin, S.S. Biochemistry, essential amino acids. In *StatPearls [Internet]*; StatPearls Publishing: St. Petersburg, FL, USA, 2024.
8. Le Couteur, D.G.; Solon-Biet, S.M.; Cogger, V.C.; Ribeiro, R.; de Cabo, R.; Raubenheimer, D.; Cooney, G.J.; Simpson, S.J. Branched chain amino acids, aging and age-related health. *Ageing Res. Rev.* **2020**, *64*, 101198. [CrossRef] [PubMed]
9. Joehnke, M.S.; Jeske, S.; Ispiryan, L.; Zannini, E.; Arendt, E.K.; Bez, J.; Sørensen, J.C.; Petersen, I.L. Nutritional and anti-nutritional properties of lentil (*Lens culinaris*) protein isolates prepared by pilot-scale processing. *Food Chem. X* **2021**, *9*, 100112.
10. Atudorei, D.; Codină, G.G. Perspectives on the use of germinated legumes in the bread making process, a review. *Appl. Sci.* **2020**, *10*, 6244. [CrossRef]
11. Criste, A.D.; Urcan, A.C.; Coroian, C.O.; Copolovici, L.; Copolovici, D.M.; Burtescu, R.F.; Oláh, N.K. Plant-Based Beverages from Germinated and Ungerminated Seeds, as a Source of Probiotics, and Bioactive Compounds with Health Benefits—Part 1: Legumes. *Agriculture* **2023**, *13*, 1185. [CrossRef]
12. Nongmaithem, R. Processing Characteristics and Utilization of Germinated Lentil for Product Development. Ph.D. Thesis, University of Saskatchewan, Saskatoon, SK, Canada, 2024.
13. Rico, D.; Peñas, E.; del Carmen García, M.; Rai, D.K.; Martínez-Villaluenga, C.; Frias, J.; Martín-Diana, A.B. Development of antioxidant and nutritious lentil (*Lens culinaris*) flour using controlled optimized germination as a bioprocess. *Foods* **2021**, *10*, 2924. [CrossRef]
14. Fouad, A.A.; Rehab, F. Effect of germination time on proximate analysis, bioactive compounds and antioxidant activity of lentil (*Lens culinaris* Medik.) sprouts. *Acta Sci. Pol. Technol. Aliment.* **2015**, *14*, 233–246. [CrossRef]
15. Horowitz, W. *AOAC Official Methods of Analysis (Sections 50.1. 21 (992.05), 45.2. 03 (944.12), and 45.2. 01 (960.46)*; Association of Official Analytical Chemists International: Gaithersburg MD, USA, 2000.
16. Jarpa-Parra, M.; Bamdad, F.; Wang, Y.; Tian, Z.; Temelli, F.; Han, J.; Chen, L. Optimization of lentil protein extraction and the influence of process pH on protein structure and functionality. *LWT-Food Sci. Technol.* **2014**, *57*, 461–469. [CrossRef]

17. Masuko, T.; Minami, A.; Iwasaki, N.; Majima, T.; Nishimura, S.-I.; Lee, Y.C. Carbohydrate analysis by a phenol–sulfuric acid method in microplate format. *Anal. Biochem.* **2005**, *339*, 69–72. [CrossRef]
18. Lee, S.C.; Prosky, L.; Vries, J.W.D. Determination of total, soluble, and insoluble dietary fiber in foods—Enzymatic-gravimetric method, MES-TRIS buffer: Collaborative study. *J. AOAC Int.* **1992**, *75*, 395–416. [CrossRef]
19. Suliman, M.A.; El Tinay, A.H.; Elkhalifa, A.E.O.; Babiker, E.E.; Elkhalil, E.A. Solubility as influenced by pH and NaCl concentration and functional properties of lentil proteins isolate. *Pak. J. Nutr.* **2006**, *5*, 589–593.
20. Marrion, O.; Fleurence, J.; Schwertz, A.; Guéant, J.-L.; Mamelouk, L.; Ksouri, J.; Villaume, C. Evaluation of protein in vitro digestibility of Palmaria palmata and Gracilaria verrucosa. *J. Appl. Phycol.* **2005**, *17*, 99–102. [CrossRef]
21. Benjakul, S.; Karnjanapratum, S.; Visessanguan, W. Hydrolysed collagen from Lates calcarifer skin: Its acute toxicity and impact on cell proliferation and collagen production of fibroblasts. *Int. J. Food Sci. Technol.* **2018**, *53*, 1871–1879. [CrossRef]
22. Ambigaipalan, P.; Shahidi, F. Date seed flour and hydrolysates affect physicochemical properties of muffin. *Food Biosci.* **2015**, *12*, 54–60. [CrossRef]
23. Wang, T.; Jónsdóttir, R.; Ólafsdóttir, G. Total phenolic compounds, radical scavenging and metal chelation of extracts from Icelandic seaweeds. *Food Chem.* **2009**, *116*, 240–248. [CrossRef]
24. Wang, B.; Wang, Y.-M.; Chi, C.-F.; Luo, H.-Y.; Deng, S.-G.; Ma, J.-Y. Isolation and characterization of collagen and antioxidant collagen peptides from scales of croceine croaker (*Pseudosciaena crocea*). *Mar. Drugs* **2013**, *11*, 4641–4661. [CrossRef]
25. Oyaizu, M. Studies on products of browning reaction antioxidative activities of products of browning reaction prepared from glucosamine. *Jpn. J. Nutr. Diet.* **1986**, *44*, 307–315. [CrossRef]
26. Karawita, R.; Siriwardhana, N.; Lee, K.-W.; Heo, M.-S.; Yeo, I.-K.; Lee, Y.-D.; Jeon, Y.-J. Reactive oxygen species scavenging, metal chelation, reducing power and lipid peroxidation inhibition properties of different solvent fractions from Hizikia fusiformis. *Eur. Food Res. Technol.* **2005**, *220*, 363–371. [CrossRef]
27. Mirzapour-Kouhdasht, A.; Moosavi-Nasab, M.; Kim, Y.-M.; Eun, J.-B. Antioxidant mechanism, antibacterial activity, and functional characterization of peptide fractions obtained from barred mackerel gelatin with a focus on application in carbonated beverages. *Food Chem.* **2021**, *342*, 128339. [CrossRef]
28. Shahidi, F.; Han, X.-Q.; Synowiecki, J. Production and characteristics of protein hydrolysates from capelin (*Mallotus villosus*). *Food Chem.* **1995**, *53*, 285–293. [CrossRef]
29. Ghumman, A.; Kaur, A.; Singh, N. Impact of germination on flour, protein and starch characteristics of lentil (*Lens culinari*) and horsegram (*Macrotyloma uniflorum* L.) lines. *LWT-Food Sci. Technol.* **2016**, *65*, 137–144. [CrossRef]
30. Barbana, C.; Boye, J.I. In vitro protein digestibility and physico-chemical properties of flours and protein concentrates from two varieties of lentil (*Lens culinaris*). *Food Funct.* **2013**, *4*, 310–321. [CrossRef]
31. Aryee, A.N.; Boye, J.I. Comparative study of the effects of processing on the nutritional, physicochemical and functional properties of lentil. *J. Food Process. Preserv.* **2017**, *41*, e12824. [CrossRef]
32. Mir, N.A.; Riar, C.S.; Singh, S. Structural modification in album (*Chenopodium album*) protein isolates due to controlled thermal modification and its relationship with protein digestibility and functionality. *Food Hydrocoll.* **2020**, *103*, 105708. [CrossRef]
33. de la Rosa-Millán, J.; Orona-Padilla, J.L.; Flores-Moreno, V.M.; Serna-Saldívar, S.O. Physicochemical, functional and ATR-FTIR molecular analysis of protein extracts derived from starchy pulses. *Int. J. Food Sci. Technol.* **2018**, *53*, 1414–1424. [CrossRef]
34. Atudorei, D.; Stroe, S.-G.; Codină, G.G. Impact of germination on the microstructural and physicochemical properties of different legume types. *Plants* **2021**, *10*, 592. [CrossRef]
35. Di, Y.; Li, X.; Chang, X.; Gu, R.; Duan, X.; Liu, F.; Liu, X.; Wang, Y. Impact of germination on structural, functional properties and in vitro protein digestibility of sesame (*Sesamum indicum* L.) protein. *LWT* **2022**, *154*, 112651. [CrossRef]
36. Setia, R.; Dai, Z.; Nickerson, M.T.; Sopiwnyk, E.; Malcolmson, L.; Ai, Y. Impacts of short-term germination on the chemical compositions, technological characteristics and nutritional quality of yellow pea and faba bean flours. *Food Res. Int.* **2019**, *122*, 263–272. [CrossRef] [PubMed]
37. Shaghaghian, S.; McClements, D.J.; Khalesi, M.; Garcia-Vaquero, M.; Mirzapour-Kouhdasht, A. Digestibility and bioavailability of plant-based proteins intended for use in meat analogues: A review. *Trends Food Sci. Technol.* **2022**, *129*, 646–656. [CrossRef]
38. Garcia-Vaquero, M.; Mora, L.; Hayes, M. In vitro and in silico approaches to generating and identifying angiotensin-converting enzyme I inhibitory peptides from green macroalga Ulva lactuca. *Mar. Drugs* **2019**, *17*, 204. [CrossRef] [PubMed]
39. Bamdad, F.; Dokhani, S.; Keramat, J.; Zareie, R. The impact of germination and in vitro digestion on the formation of angiotensin converting enzyme (ACE) inhibitory peptides from lentil proteins compared to whey proteins. *Heart Fail* **2009**, *3*, 109–119.
40. Ratnayani, K.; Suter, I.K.; Antara, N.S.; Putra, I.N.K. Angiotensin converting enzyme (ACE) inhibitory activity of peptide fraction of germinated Pigeon Pea (*Cajanus cajan* (L.) Millsp.). *Indones. J. Chem.* **2019**, *19*, 900–906. [CrossRef]
41. Rezvankhah, A.; Yarmand, M.S.; Ghanbarzadeh, B.; Mirzaee, H. Generation of bioactive peptides from lentil protein: Degree of hydrolysis, antioxidant activity, phenol content, ACE-inhibitory activity, molecular weight, sensory, and functional properties. *J. Food Meas. Charact.* **2021**, *15*, 5021–5035. [CrossRef]

42. Boye, J.I.; Roufik, S.; Pesta, N.; Barbana, C. Angiotensin I-converting enzyme inhibitory properties and SDS-PAGE of red lentil protein hydrolysates. *LWT-Food Sci. Technol.* **2010**, *43*, 987–991. [CrossRef]
43. Hanafi, M.A.; Hashim, S.N.; Chay, S.Y.; Ebrahimpour, A.; Zarei, M.; Muhammad, K.; Abdul-Hamid, A.; Saari, N. High angiotensin-I converting enzyme (ACE) inhibitory activity of Alcalase-digested green soybean (*Glycine max*) hydrolysates. *Food Res. Int.* **2018**, *106*, 589–597. [CrossRef]
44. Boschin, G.; Scigliuolo, G.M.; Resta, D.; Arnoldi, A. ACE-inhibitory activity of enzymatic protein hydrolysates from lupin and other legumes. *Food Chem.* **2014**, *145*, 34–40. [CrossRef]
45. Hlosrichok, A.; Aunpad, R. Antioxidant potential of gamma ray irradiated winged bean (*Psophocarpus tetragonolobus*) seed protein hydrolysate. *Funct. Foods Health Dis.* **2023**, *13*, 208–224. [CrossRef]
46. Ribeiro, A.C.; Ferreira, R.; Freitas, R. Plant lectins: Bioactivities and bioapplications. *Stud. Nat. Prod. Chem.* **2018**, *58*, 1–42.
47. Balanescu, F.; Busuioc, A.C.; Botezatu, A.V.D.; Gosav, S.; Avramescu, S.M.; Furdui, B.; Dinica, R.M. Comparative Study of Natural Antioxidants from *Glycine max, Anethum graveolensand Pimpinella anisum* Seed and Sprout Extracts Obtained by Ultrasound-Assisted Extraction. *Separations* **2022**, *9*, 152. [CrossRef]
48. Olagunju, A.I.; Omoba, O.S.; Enujiugha, V.N.; Alashi, A.M.; Aluko, R.E. Pigeon pea enzymatic protein hydrolysates and ultrafiltration peptide fractions as potential sources of antioxidant peptides: An in vitro study. *LWT* **2018**, *97*, 269–278. [CrossRef]
49. Chaijan, M.; Panpipat, W. Nutritional composition and bioactivity of germinated Thai indigenous rice extracts: A feasibility study. *PLoS ONE* **2020**, *15*, e0237844. [CrossRef] [PubMed]
50. Parit, S.B.; Dawkar, V.V.; Tanpure, R.S.; Pai, S.R.; Chougale, A.D. Nutritional quality and antioxidant activity of wheatgrass (*Triticum aestivum*) unwrap by proteome profiling and DPPH and FRAP assays. *J. Food Sci.* **2018**, *83*, 2127–2139. [CrossRef]
51. Popoola, O.O. Phenolic compounds composition and in vitro antioxidant activity of Nigerian Amaranthus viridis seed as affected by autoclaving and germination. *Meas. Food* **2022**, *6*, 100028. [CrossRef]
52. Martínez, K.D.; Sánchez, C.C.; Patino, J.M.R.; Pilosof, A.M. Interfacial and foaming properties of soy protein and their hydrolysates. *Food Hydrocoll.* **2009**, *23*, 2149–2157. [CrossRef]
53. Rao, G.N.; Balaswamy, K.; Rao, P.P.; Rao, D.G.; Satyanarayana, A. Nutritional quality, fatty acids, amino acids and functional characteristics of bael (*Aegle marmelos* L.) seed protein concentrate. *J. Food Drug Anal.* **2011**, *19*, 528–542.
54. Liu, M.; Childs, M.; Loos, M.; Taylor, A.; Smart, L.B.; Abbaspourrad, A. The effects of germination on the composition and functional properties of hemp seed protein isolate. *Food Hydrocoll.* **2023**, *134*, 108085. [CrossRef]
55. Martinez, K.; Baeza, R.; Millan, F.; Pilosof, A. Effect of limited hydrolysis of sunflower protein on the interactions with polysaccharides in foams. *Food Hydrocoll.* **2005**, *19*, 361–369. [CrossRef]
56. Vogelsang-O'Dwyer, M.; Sahin, A.W.; Arendt, E.K.; Zannini, E. Enzymatic hydrolysis of pulse proteins as a tool to improve techno-functional properties. *Foods* **2022**, *11*, 1307. [CrossRef]
57. Gharibzahedi, S.M.T.; Smith, B. The functional modification of legume proteins by ultrasonication: A review. *Trends Food Sci. Technol.* **2020**, *98*, 107–116. [CrossRef]

foods

MDPI

Article

Effects of Conjugation with Basil Seed Gum on Physicochemical, Functional, Foaming, and Emulsifying Properties of Albumin, Whey Protein Isolate and Soy Protein Isolate

Hadi Hashemi [1], Mohammad Hadi Eskandari [1], Mohammadreza Khalesi [2], Mohammad-Taghi Golmakani [1], Mehrdad Niakousari [1] and Seyed Mohammad Hashem Hosseini [1,*]

[1] Department of Food Science and Technology, School of Agriculture, Shiraz University, Shiraz 71441-13131, Iran; hadihashemigahruie@shirazu.ac.ir (H.H.); eskandar@shirazu.ac.ir (M.H.E.); golmakani@shirazu.ac.ir (M.-T.G.); niakosar@shirazu.ac.ir (M.N.)

[2] Department of Biological Sciences, University of Limerick, V94 T9PX Limerick, Ireland; mohammadreza.khalesi@ul.ie

* Correspondence: hhosseini@shirazu.ac.ir

Abstract: Protein conjugation with the Maillard reaction has received considerable attention in the past decades in terms of improving functional properties. This study evaluated the changes in the techno-functional properties of whey protein isolate (WPI), soy protein isolate (SPI), and albumin (Alb) after conjugation with basil seed gum (BSG). The conjugates were developed via the Maillard reaction. Various analyses including FT-IR, XRD, SEM, SDS-PAGE, DSC, RVA, rheology, zeta potential, emulsion, and foaming ability were used for evaluating conjugation products. Conjugation between proteins (WPI, SPI, Alb) and BSG was validated by FT-IR spectroscopy. XRD results revealed a decrease in the peak of BSG after conjugation with proteins. SDS-PAGE demonstrated the conjugation of WPI, SPI, and Alb with BSG. DSC results showed that conjugation with BSG reduced the T_g of WPI, SPI, and Alb from 210.21, 207.21, and 210.90 °C to 190.30, 192.91, and 196.66 °C, respectively. The emulsion activity and emulsion stability of protein/BSG conjugates were increased significantly. The droplet size of emulsion samples ranged from 112.1 to 239.3 nm on day 3. Nanoemulsions stabilized by Alb/BSG conjugate had the smallest droplet sizes (112.1 and 143.3 nm after 3 and 17 days, respectively). The foaming capacity of WPI (78.57%), SPI (61.91%), and Alb (71.43%) in their mixtures with BSG increased to 107.14%, 85.71%, and 85.71%, respectively, after making conjugates with BSG. The foam stability of WPI (39.34%), SPI (61.57%), and Alb (53.37%) in their mixtures with BSG (non-conjugated condition) increased to 77.86%, 77.91%, and 72.32%, respectively, after formation of conjugates with BSG. Conjugation of BSG to proteins can improve the BSG applications as a multifunctional stabilizer in pharmaceutical and food industries.

Keywords: albumin; basil seed gum; conjugation; soy protein isolate; whey protein isolate

1. Introduction

Ocimum basilicum, from the Laminaceae family, is grown in different locations, mainly in France, Italy, Egypt, Bulgaria, South Africa, Hungary, and North America [1]. Because of its pleasant and specific flavor characteristics, basil is usually used for food production. Also, basil seed is popular due to many health benefits such as relieving ulcer, cough, diarrhea, kidney disorder, indigestion, and sore throat. Therefore, it is a popular ingredient in traditional medicine [2]. The seeds' mucilage, isolated by basil seed hydration in water, contains high amount of polysaccharides [3]. Basil seed gum (BSG) has the potential to create functional characteristics including emulsion and foam stabilization, thickening, mouthfeel improvements, and edible coating [4,5].

Improving the functional characteristics of proteins is important for the producers of food, cosmetic, and pharmaceutical products. A lot of modifications including physical, chemical, or enzymatic techniques have been applied to increase proteins' functionality [6–12]. Chemical modifications due to the formation of detrimental products or potential health hazards have not been extensively used for food production. Therefore, a different strategy is needed to increase the proteins' functional properties in the food industry.

Today, research on the polysaccharide–protein conjugation reaction, based on the Maillard-type reactions, which includes the reducing end carbonyl groups of polysaccharides and the protein's free amino groups, have been increased. This chemical modification can improve the product's functional properties [13–15]. As an example, the emulsifying properties of proteins are significantly improved by the Maillard reaction [16]. The main aspect of the prepared protein/polysaccharide conjugates is the higher foaming and emulsifying capability, which is practical for food industry [16,17]. Also, conjugation can improve the solubility of proteins, and the antioxidant and antibacterial properties of polysaccharides and proteins [11,18]. The allergenicity of some proteins is reduced by its conjugation with polysaccharides [19,20]. Furthermore, the comprehensive review by Urango et al. [21] on Maillard conjugates in health showed that the products of the Maillard reaction can modulate the intestinal microbiota composition in adolescents during in vivo testing. It can also increase immunomodulatory effects, improve immunity, increase the production of various immune cells, improve the bioaccessibility of total carotenoids, increase anti-inflammatory effects, and inhibit cell proliferation activity in human breast cancer cells.

Research on the application of protein/polysaccharides conjugates as emulsifying and foaming agents has increased. Improvement in the foam and emulsion properties by Maillard conjugates were reported in several studies in 2024 by Zhang et al. [22] on the SPI hydrolysate-various polysaccharides (e.g., gum Arabic, sodium alginate, xanthan gum, and soybean polysaccharides) conjugates, Tao et al. [23] on the SPI/soybean peptide and ginseng polysaccharide conjugates, and Amiratashani et al. [24] on the grass pea protein and xanthan gum conjugates.

This technique uses a naturally occurring reaction without the need of chemical solvents. So, it can be considered an important method for industrial application. To our knowledge, there is no report on the conjugation of whey protein isolate (WPI), soy protein isolate (SPI), and albumin (Alb) with BSG. Therefore, the aim of this study was to evaluate the changes in the physicochemical, functional, thermal, foaming, and emulsifying properties of WPI, SPI, or Alb after fabricating conjugates with BSG using the dry method under controlled temperature and relative humidity conditions.

2. Materials and Methods

2.1. Materials

BSG (moisture (5.74%), protein (1.28%), and carbohydrate (84.74%)) were obtained from Reyhan Gum Parsian (Tehran, Iran). WPI (protein > 90%), SPI (protein > 92%), and Alb (protein > 98%) were obtained from Davisco Foods International Inc. (Eden Prairie, MN, USA), G max Chemicals Co., Ltd. (Shandong, China), and Sigma Chemical Co. (St. Louis, MO, USA), respectively. Canola oil was purchased from Narges Oil Co. (Shiraz, Iran). Other reagents and materials were of analytical grade.

2.2. Preparation of BSG Conjugates

Proteins (100 mg) and BSG (100 mg) were mixed and dispersed in 10 mL sodium phosphate buffer (0.05 M, pH = 8.5 ± 0.1). After overnight hydration, the dispersions were lyophilized and milled into a fine powder. Powdered samples were kept in sealed glass desiccators containing saturated KBr solution (RH ~ 79%) at the bottom. Desiccators were kept in an oven at 60 °C. The conjugates (BSG/WPI conjugate, BSG/SPI conjugate, and BSG/Alb conjugate) were harvested after 10 days and evaluated [25]. Control samples (BSG/WPI mixture, BSG/SPI mixture, BSG/Alb mixture) containing non-conjugated mixture of BSG and three different types of proteins were also investigated.

2.3. FT-IR Spectroscopy

An FT-IR spectrometer (Avatar 370, Thermo Nicolet Corp., Madison, WI, USA) was used to study the functional groups and structural characteristics of proteins/BSG mixtures or conjugates. All samples were mixed with KBr to prepare pellets. The spectra were recorded from 4000 to 400 cm^{-1} [24].

2.4. Microstructure (SEM)

The surface microstructure of mixtures and conjugates were studied using a scanning electron microscope (TESCAN vega3, Brno, Czech Republic). The samples were placed on an aluminum tape and sputter-coated with gold (Desk Sputter Coater DSR1, Nanostructural Coating Co., Isfahan, Iran) [24].

2.5. X-Ray Diffraction

An X-ray diffraction analysis of proteins/gum mixtures or conjugates was carried out by a Bruker ASX D8 X-ray powder diffractometer (Bruker, Billerica, MA, USA) with Cu Ka radiation operated at 40 kV and 40 mA. Analysis was performed between 5° and 70° (2θ) with 0.02° step.

2.6. Sodium Dodecyl Sulfate Polyacrylamide Gel Electrophoresis (SDS-PAGE)

Slab SDS-PAGE was settled based on the discontinuous buffer system of Laemmli [26]. Mixtures and conjugates were mixed with a loading buffer to create a final concentration of 1 mg/mL. Electrophoresis was performed at 25 mA and the gel was stained with Coommassie Brilliant blue R-250 (0.25%).

2.7. Zeta Potential

The zeta potential of mixtures and conjugates (at 0.1% *w*/*w* concentration) was measured by a Zetasizer (SZ100, Horiba, Japan).

2.8. Differential Scanning Calorimetry (DSC) and Thermogravimetric Analysis (TGA)

Almost 5 mg of mixtures and conjugates were analyzed by a differential scanning calorimeter and Thermogravimetric analysis (TGA) (Perkin-Elmer, Beaconsfield, UK). Analysis was performed in the temperature range of 30–230 °C under nitrogen flow and with a heating rate of 10 °C/min [27].

2.9. Temperature Effects on Viscosity

The effect of temperature on the viscosity of mixtures and conjugates was studied by a Rapid Visco Analyser (RVA Starch Master 2, Perten, Australia). Various dispersions (25 mL) of mixtures and conjugates were prepared at 0.3% *w*/*v* and then analyzed. The changes in viscosity were monitored by heating from 50 to 95 °C and, then by cooling from 95 to 50 °C. The paddle rotation speed was 160 rpm [28].

2.10. Apparent Viscosity Measurement

The effect of shear rate on the apparent viscosity of mixtures and conjugates was determined by a Brookfield viscometer (DV2 Pro II, Brookfield Engineering Laboratories, Middleboro, MA, USA) equipped with spindle cp51. The commonly used models including Power law, Bingham, and Casson were used for data fitting [28].

2.11. Emulsion Preparation

Protein/BSG mixtures or conjugates (0.075 g) were dispersed in double distilled water (DDW) (17.50 g) and hydrated at 25 °C for 24 h. After that, canola oil (7.50 g) was gradually added, and homogenized by Ultra-Turrax (T18, IKA, Staufen im Breisgau, Germany) at 15,000 rpm for 2 min, followed by sonication (SONOPULS, HD3200, Bandelin Co., Berlin, Germany) for 5 min at 150 W.

2.11.1. Emulsion Activity

The emulsion activity index (EAI) of protein/BSG mixtures or conjugates was measured according to the literature [28]. Emulsion samples (100 mL) were diluted in DDW (25 mL) and absorption was measured at 500 nm. EAI was determined by Equation (1):

$$EAI\ (\mathrm{m^2/g}) = (2 \times 2.303 \times D \times A)/(I \times \Phi \times C \times 10{,}000), \qquad (1)$$

where A is the absorbance at 500 nm, I is the optical path length of the cuvette (m), D is the dilution factor, Φ is the oil volume fraction, and C is the amount of emulsifier in aqueous phase (g/cm^3).

2.11.2. Emulsion Stability

Emulsions were poured into in 15 mL PE tubes, capped and stored at room temperature. The gravitational stability of emulsions was investigated after 1, 3, 7, 10, and 17 days according to the method described by Kheynoor et al. [29].

2.11.3. Droplet Size

The droplet size distribution of nanoemulsions was determined by a dynamic light scattering (DLS) technique (SZ100, Horiba, Japan). At first, emulsion samples were diluted 100 times with DDW. Then, volume weighted diameter ($D_{4,3}$) and distribution width (Span) were analyzed over time [30].

2.11.4. Zeta Potential

The zeta potential of samples was evaluated over time by a DLS instrument (SZ100, Horiba, Japan). Samples were diluted 100 times with DDW before measurement. The electrophoretic mobility of the samples was transformed to zeta potential using the Smoluchowski equation [31].

2.12. Determination of Foaming Capacity and Foam Stability

The foaming capacity and foam stability were measured at a 0.3% (*w*/*w*) concentration based on the technique reported by Naji-Tabasi and Razavi [5] with some modifications. The dispersion (20 mL) of mixtures and conjugates (0.6% (*w*/*v*) was prepared and kept for 24 h under mixing, followed by whipping vigorously at 15,000 rpm for 2 min using a homogenizer (Ultra Turrax T18, Staufen im Breisgau, Germany).

2.13. Statistical Analysis

The results obtained from all tests were analyzed using one-way analysis of variance (ANOVA) at $p < 0.05$ significance. Duncan's multiple range tests were applied by SAS® software (ver. 9.1, SAS Institute Inc., Cary, NC, USA) to determine significant differences.

3. Results and Discussion

3.1. FT-IR Analysis

FT-IR spectroscopy is an important technique for evaluating protein–polysaccharide covalent interactions, since most characteristics regions related to both biopolymers do not overlap [32]. Figure 1 illustrates the FT-IR spectra of different protein/BSG mixtures and protein/BSG conjugates. The amide I and amide II bands, which are observed between 1650 cm^{-1} and 1540 cm^{-1}, respectively, are the most specific spectral properties of proteins [33]. The main changes after conjugation were observed at around 850 to 1150 cm^{-1}. The peaks at 1653.26–1655.67 and 1542.85–1545.39 cm^{-1} were attributed to amide I (C=O stretching) and amide II (N-H bending), respectively (Table 1). The band around 3420 cm^{-1} was related to the inter- and intra-molecular hydrogen bonds. This band is broad due to the overlapping of OH and -NH bands. The strong band between ~2980 and 3100 cm^{-1} is assigned to C-H stretching vibration [34]. The C=C (1653 cm^{-1}) and C=N (1402 cm^{-1}) stretching vibrations are the main bands in BSG. The C=O stretching was observed at 1055 cm^{-1} [35,36].

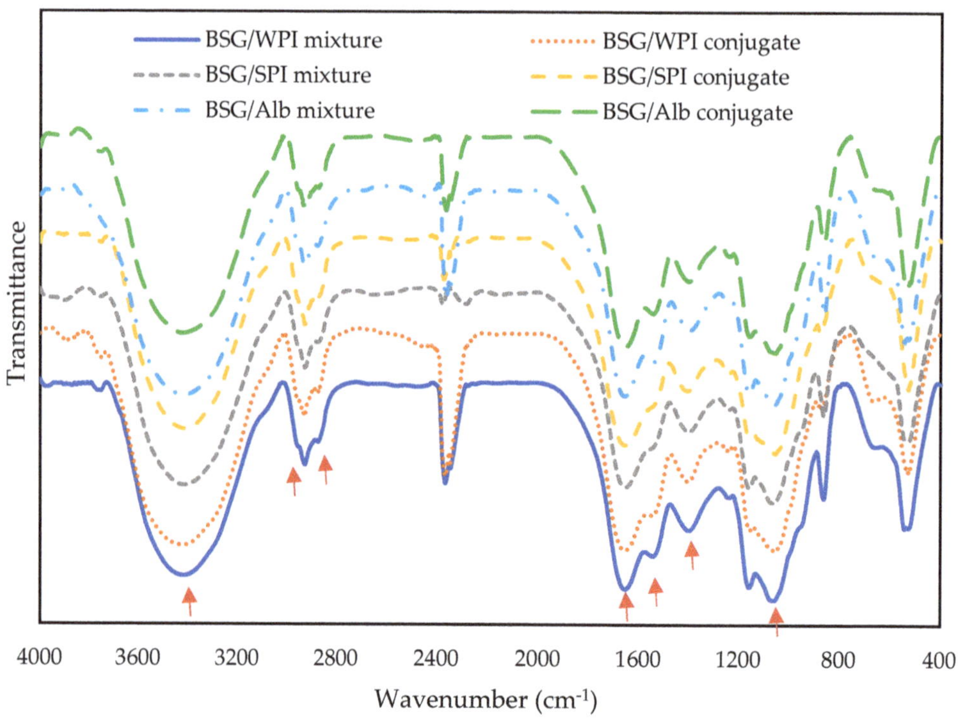

Figure 1. FT-IR spectra of BSG/WPI mixture, BSG/WPI conjugate, BSG/SPI mixture, BSG/SPI conjugate, BSG/Alb mixture, and BSG/Alb conjugate. Red arrows are the main peaks.

Table 1. FT-IR peak assignment of BSG/WPI mixture, BSG/WPI conjugate, BSG/SPI mixture, BSG/SPI conjugate, BSG/Alb mixture, and BSG/Alb conjugate.

Assignment	Samples' Wavenumber (cm^{-1})					
	BSG/WPI Mixture	BSG/WPI Conjugate	BSG/SPI Mixture	BSG/SPI Conjugate	BSG/Alb Mixture	BSG/Alb Conjugate
-OH stretching vibrations	3420	3424	3417	3412	3418	3425
$-CH_2-$ and >CH- stretching and bending vibrations	2929	2929	2928	2931	2928	2928
C-H absorption	2874	2874	2878	2873	2878	2870
C=O stretching as amide I	1653	1654	1653	1654	1655	1654
-NH twisting (amide II)	1545	1542	1540	1544	1544	1545
C=N stretching vibrations	1402	1406	1402	1408	1400	1402
-OH bending vibration	1238	1240	1238	1236	1236	1235
C-O-C stretching vibrations of glycosidic bonds	1060	1055	1060	1055	1060	1055

Protein/polysaccharide conjugation decreases the peak intensity of NH_2 functional groups and increases the bands related to Amadori ingredients (C=O) and Schiff base (C=N) [37]. The fluctuations in the band intensity of C-O and C-N functional groups (at 1653 and 1545 cm^{-1}, respectively) might reflect the conjugation phenomenon, and thus, the changes in the secondary structure of proteins [38]. When a protein/polysaccharide mixture is heated under certain conditions, the bands intensity at 1545 and 1055 cm^{-1} is increased [11] and a possible new band at 2870 cm^{-1} is observed. These changes are due to the formation of a Schiff base (C=N double bond) during the reaction between carbonyl groups of polysaccharides and amino groups of proteins. C-N stretching in the Schiff base appears at 1610–645 cm^{-1} depending on the ligand type [38].

3.2. Microstructure Properties

Scanning electron microscopy is usually applied to study the microstructure of polysaccharides/proteins mixtures and conjugates [39]. Changes in the microstructure during conjugation are finally reflected in the techno-functional properties of conjugates. The SEM micrographs of mixtures and conjugates are shown in Figure 2 (or Figure S1). Aggregation or association was not observed in non-conjugated samples (Figure 2a,c,e). However, in conjugated samples (Figure 2b,d,f), the surface morphology revealed close associations and more compact microstructure. Niu et al. [40] found changes in the wheat germ protein (WGP) surface structures upon conjugation with dextran. They reported that the WGP particles were large, irregular and uneven, while WGP/dextran conjugates were thin sheets. A similar observation of conjugated microstructure was reported by Yadav et al. [41] for the mixtures and conjugates of milk protein/saccharide.

3.3. X-Ray Diffraction

X-ray diffraction was applied to evaluate the effect of conjugation on the amorphous state of biopolymers. As illustrated in Figure 3, various protein/BSG mixtures and conjugates exhibited fairly similar diffraction patterns. After mixing BSG with WPI, SPI, and Alb, the mixture showed broader peaks at 2θ values of 12.9° and 32.1°, 13.4° and 32°, as well as 13.2° and 32.1°, respectively. For the conjugates of BSG with WPI, SPI, and Alb, the position of crystalline peaks appeared at 2θ values of 13.3° and 31.7°, 13.7° and 31.8°, as well as 13.95° and 31.7°, respectively. These results showed that the conjugation of proteins with BSG does not affect the amorphous state of biopolymers. Slight changes in peak intensity were observed at 2θ around 13°. Similar results were reported in conjugation between gum Arabic and canola protein isolate [42].

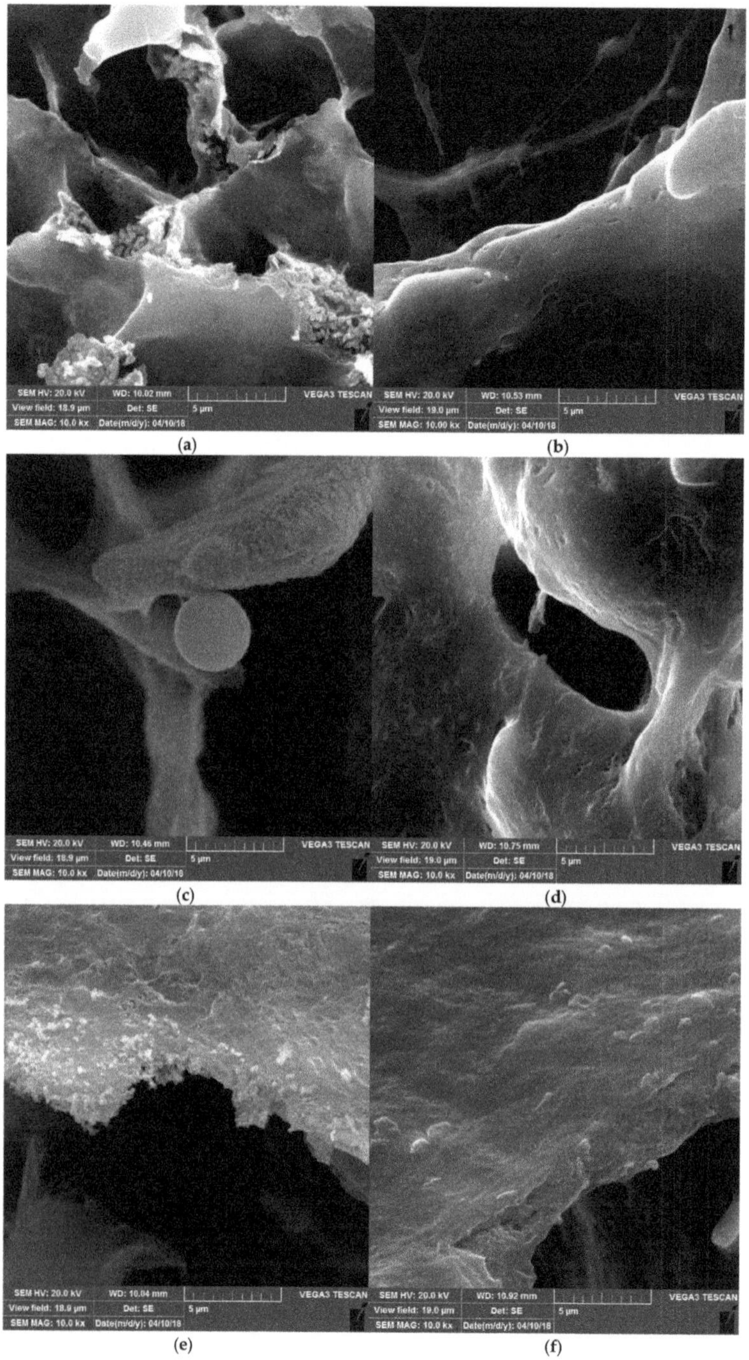

Figure 2. Scanning electron micrographs of (**a**) BSG/WPI mixture, (**b**) BSG/WPI conjugate, (**c**) BSG/SPI mixture, (**d**) BSG/SPI conjugate, (**e**) BSG/Alb mixture, and (**f**) BSG/Alb conjugate (10 days, T = 60 °C at pH = 8.5). Magnification 10,000×; scale bar = 5 μm.

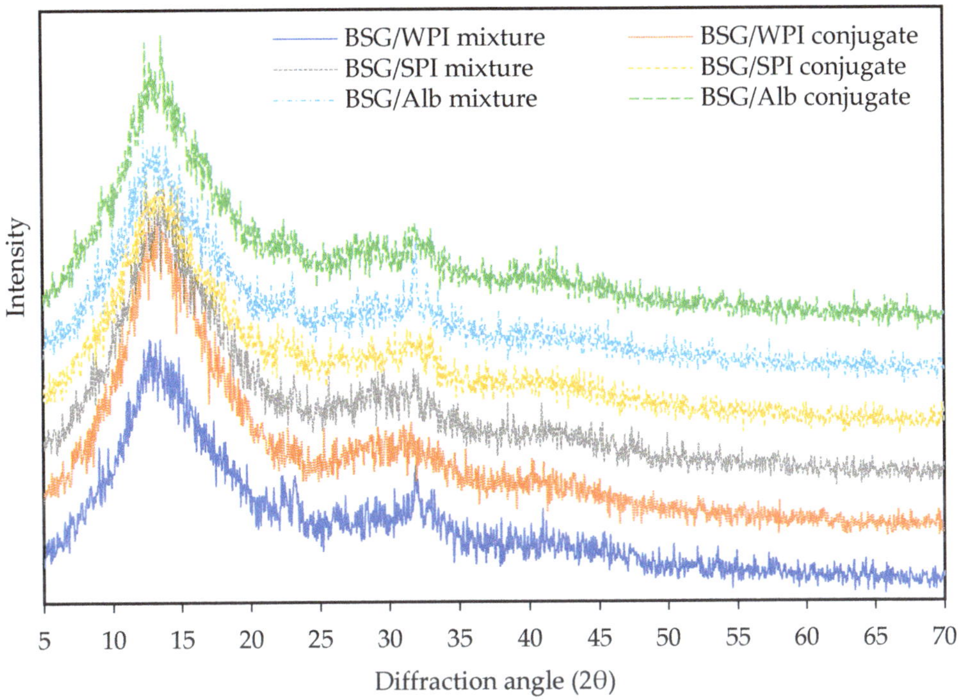

Figure 3. X-ray diffraction patterns of various proteins/BSG mixtures and conjugates.

3.4. Protein Analysis (SDS-PAGE)

The SDS-PAGE patterns of mixtures and conjugates are shown in Figure 4. The bands of BSG/WPI conjugate, BSG/SPI conjugate and BSG/Alb conjugate were sharper than those of BSG/WPI mixture, BSG/SPI mixture, and BSG/Alb mixture because of the conjugation reaction, as it caused higher molecular weight. The sharper bands observed in SDS-PAGE image probably originated from the addition of a considerable amount of BSG molecules to WPI, SPI, and Alb, and thus changes in protein conformation, disulfide-sulfhydryl bands, and interaction of protein molecules with each other [43]. The variation in band intensity was reported in many studies during the conjugation of various polysaccharides with LZM [44], milk proteins [41], and soy protein isolate [45]. Similarly, in this study, the SDS-PAGE confirmed that the conjugation of WPI, SPI, and Alb with BSG caused the formation of molecules with higher molecular weights.

3.5. Differential Scanning Calorimetry (DSC) and Thermogravimetric Analysis (TGA)

Protein denaturation changes the techno-functional properties. The thermodynamic stability of proteins is usually determined by the DSC technique [46]. This technique is also applied for evaluating the thermal properties of protein–polysaccharide complexes [47]. Table 2 and Figure 5 show the thermal characteristics of BSG/protein mixtures and respective conjugates. The results showed that the BSG/WPI conjugate, BSG/SPI conjugate, and BSG/Alb conjugate had lower onset (T_{onset}), peak ($T_{midpoint}$), and endset (T_{endset}) temperatures than their mixtures. However, the denaturation enthalpy (ΔH) of mixtures was higher than that of conjugates. After the conjugation of WPI, SPI, and Alb with BSG, the enthalpy reduced from −52.47, −124.66 and −67.07 J/g to −203.36, −336.31 and −80.82 J/g, respectively. This reduction might be due to the aggregation and disruption of hydrophobic interactions during the exothermic reactions [48]. Protein denaturation is one of the main parameters for changing the enthalpy [47]. Takahashi et al. [49] studied the conjugation of LZM with glucose stearic acid. They reported that denaturation temperature was increased and ΔH was decreased due to the change in the α-helix in LZM and to the molecular interaction and aggregation.

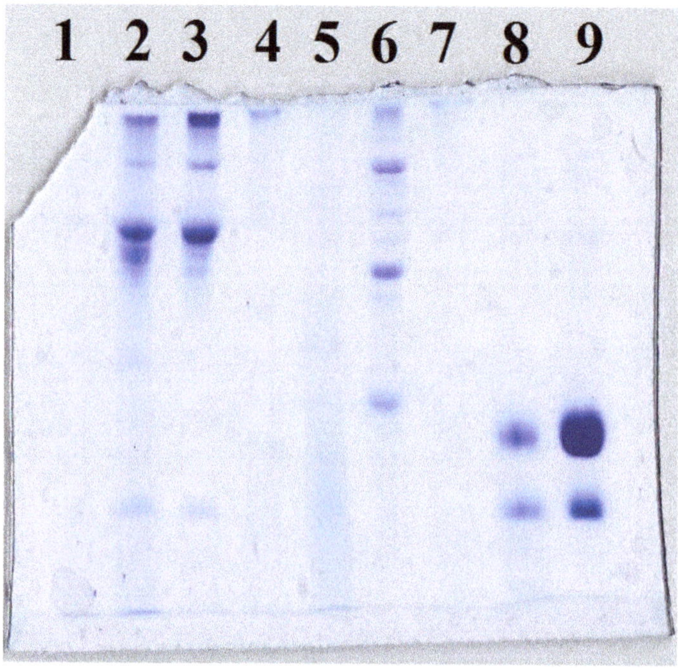

Figure 4. SDS-PAGE patterns of (2) BSG/WPI mixture, (3) BSG/WPI conjugate, (5) BSG/SPI mixture, (6) BSG/SPI conjugate, (8) BSG/Alb mixture, and (9) BSG/Alb conjugate. (1), (4) and (7) are hole without any sample.

Table 2. Thermal properties of mixture and conjugate of WPI, SPI, and Alb with BSG as evaluated by DSC.

Samples	Denaturation Temperature (°C)			
	T_{oneset}	Tg	T_{endset}	ΔH (J/g)
BSG/WPI conjugate	198.12	210.21	222.60	−52.47
BSG/WPI mixture	164.06	190.30	217.35	−203.36
BSG/SPI conjugate	182.18	207.21	226.73	−124.66
BSG/SPI mixture	162.76	192.91	222.52	−336.31
BSG/Alb conjugate	185.65	210.90	224.86	−67.07
BSG/Alb mixture	176.77	196.66	219.57	−80.82

Thermogravimetric analysis (TGA) showed two main steps of weight loss. The first weight loss < 90 °C was because of the evaporation of water, and the second weight loss (from 180 to 230 °C) was due to the thermal decomposition of hydrocolloids [50].

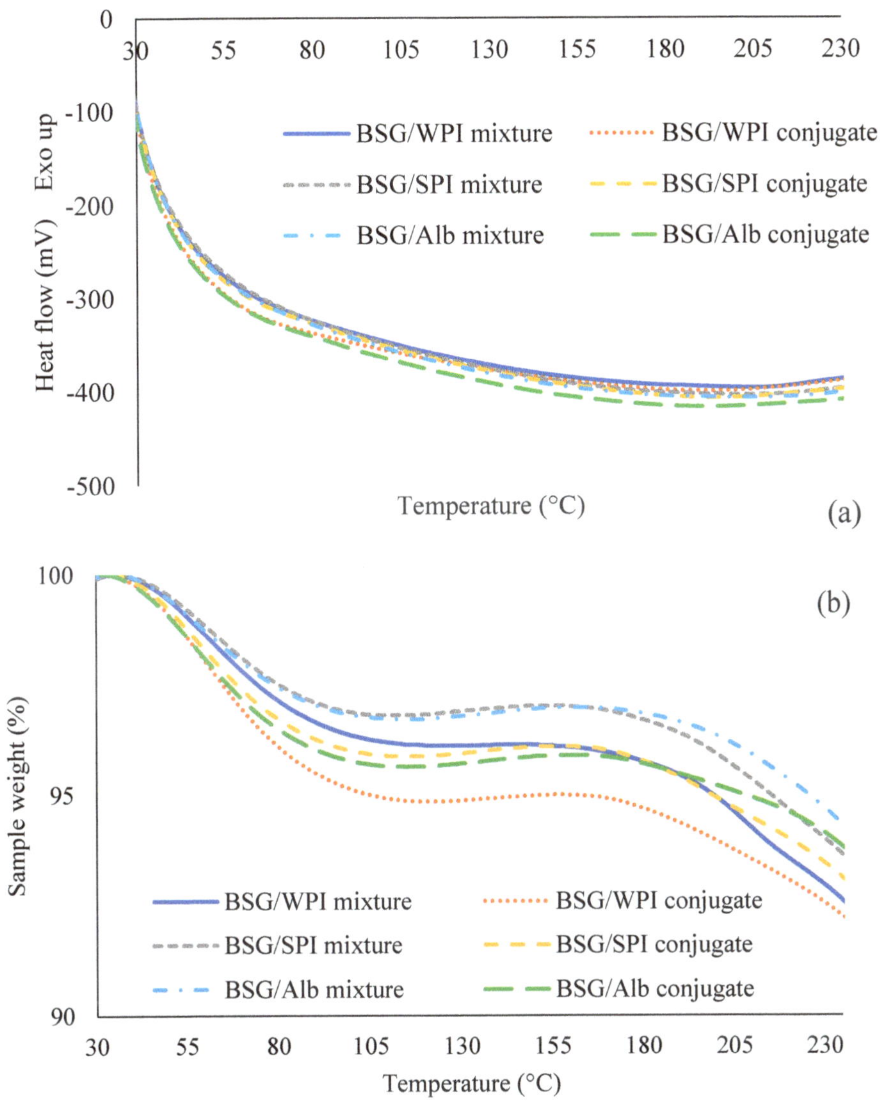

Figure 5. DSC thermograms (**a**) and TGA curve (**b**) of BSG/WPI mixture, BSG/WPI conjugate, BSG/SPI mixture, BSG/SPI conjugate, BSG/Alb mixture, and BSG/Alb conjugate.

3.6. Rheological Results

The effect of temperature on the viscosity of different BSG/protein mixtures and BSG/protein conjugates (0.3% *w*/*v*) was measured by RVA. The findings are shown in Figure 6. The results indicated that in some samples, the rise in temperature to 95 °C is accompanied by increasing in viscosity, which could be attributed to the higher hydration induced by a higher temperature. The peak viscosity values of the BSG/WPI mixture, BSG/WPI conjugate, BSG/SPI mixture, BSG/SPI conjugate, BSG/Alb mixture, and BSG/Alb conjugate solution were 384, 192, 186, 178, 759, and 428 cp, respectively. Generally, the viscosity of the BSG/protein mixture was higher than that of the respective

BSG/protein conjugate. This might be due to the changes in the excluded volume of biopolymers after the formation of conjugates.

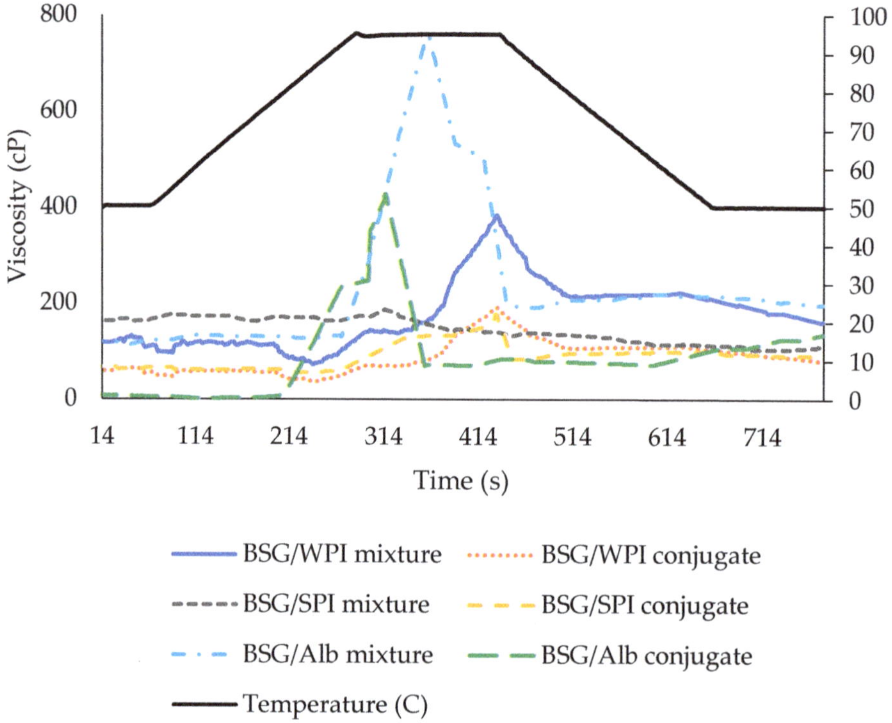

Figure 6. Instrumental rheological performance of BSG/WPI mixture, BSG/WPI conjugate, BSG/SPI mixture, BSG/SPI conjugate, BSG/Alb mixture, and BSG/Alb conjugate employing Rapid Visco Analyser.

All samples showed shear-thinning behavior (i.e., a decrease in viscosity by increasing shear rate (Figure 7)). High MW, aggregation, and disentanglement of polymers under shear are the main reasons for shear-thinning behavior [51]. Similar results have already been reported [4]. The results of data fitting with Power Law, Bingham, and Casson models were reported in Table 3. All models exhibited high R^2 values (0.957–0.995). The value of flow behavior index (*n*) of Power Law model (0.01–0.26) also confirmed the shear-thinning behavior ($n < 1$) of mixtures and conjugates of BSG with WPI, SPI, and Alb (Table 3). Shear-thinning hydrocolloids are utilized to change the viscosity of foods. Taking into account the recovery after removing shear, this feature is useful during some unit operations including filling and pumping [52]. The consistency coefficient (*k*) of samples ranged from 449 to 967.8 mPa s^n. The results showed that the conjugation of BSG with WPI, SPI, and Alb did not have a significant effect on consistency coefficient (Table 3).

Table 3. Viscosity steady shear rheological parameters of solutions (20 °C).

		Bingham			Casson			Power Law		
	Vis (cP) at 50.3 1/s	**K (Pa s^n)**	**$\tau_{0\ (Pa)}$**	**R^2**	**K (Pa s^n)**	**$\tau_{0\ (Pa)}$**	**R^2**	**K (mPa s^n)**	**n**	**R^2**
BSG/WPI mixture	29.27 ± 3.67 *	4.95 ± 0.72	10.24 ± 2.56	96.73 ± 1.86	1.28 ± 0.17	7.80 ± 2.28	98.73 ± 0.42	536.10 ± 185.12	0.23 ± 0.04	97.60 ± 0.56
BSG/WPI conjugate	16.30 ± 1.57	2.14 ± 0.90	7.06 ± 2.30	96.83 ± 3.50	0.69 ± 0.01	3.36 ± 1.16	98.50 ± 1.68	449.00 ± 352.26	0.26 ± 0.16	96.97 ± 3.50
BSG/SPI mixture	23.53 ± 1.83	2.96 ± 0.18	15.38 ± 8.70	95.70 ± 2.95	0.53 ± 0.09	8.64 ± 0.53	98.20 ± 1.25	646.20 ± 4.67	0.17 ± 0.02	96.67 ± 1.76
BSG/SPI conjugate	17.97 ± 2.88	0.38 ± 0.11	7.42 ± 2.83	99.13 ± 0.29	0.02 ± 0.01	7.21 ± 2.85	99.57 ± 0.15	686.97 ± 287.90	0.03 ± 0.02	99.07 ± 0.32
BSG/Alb mixture	19.64 ± 2.25	1.41 ± 0.62	9.16 ± 1.26	99.00 ± 0.40	0.07 ± 0.03	8.44 ± 1.35	99.27 ± 0.35	746.20 ± 147.77	0.08 ± 0.04	98.00 ± 0.95
BSG/Alb conjugate	18.28 ± 6.59	0.05 ± 0.05	9.43 ± 3.46	99.07 ± 0.38	0.01 ± 0.01	9.51 ± 3.61	99.50 ± 0.26	967.87 ± 390.47	0.01 ± 0.01	98.90 ± 0.56

* Data represent mean ± standard deviation of three independent repeats.

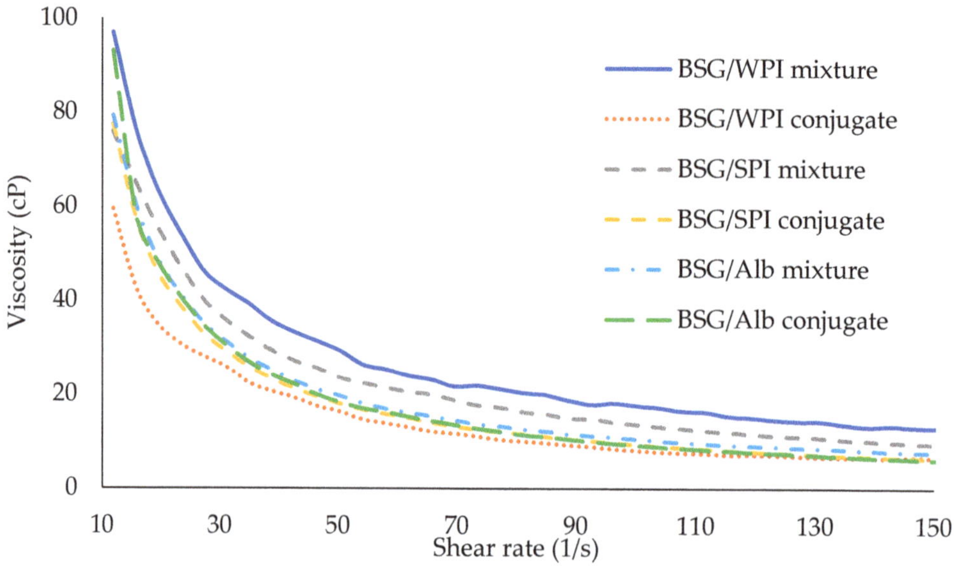

Figure 7. Changes in the apparent viscosity of BSG/WPI mixture, BSG/WPI conjugate, BSG/SPI mixture, BSG/SPI conjugate, BSG/Alb mixture, and BSG/Alb as a function of shear rate (0.3% w/w, 20 °C).

3.7. Zeta Potential

The zeta potential shows the surface charge at the distance of the slipping plane of hydrochlorides and indicates the electrostatic repulsion among charged molecules. Table 4 reports the zeta potential values of BSG/protein mixtures and BSG/protein conjugates. The BSG/WPI mixture, BSG/SPI mixture, and BSG/Alb mixture had negative zeta potential values of −39.50, −41.60, and −43.83 mV, respectively. The negative values were attributed to the pH values above proteins' isoelectric point and negative charges along BSG backbone. Conjugation increased the absolute values of zeta potential to −45.43, −45.30, and −51.97 for BSG/WPI conjugate, BSG/SPI conjugate, and BSG/Alb conjugate, respectively. The increase in zeta potential absolute values might be attributed to the blocking of positively charged amino groups of proteins during conjugation.

Table 4. Zeta potential values of BSG/protein mixtures and BSG/protein conjugates.

	BSG/WPI Mixture	BSG/WPI Conjugate	BSG/SPI Mixture	BSG/SPI Conjugate	BSG/Alb Mixture	BSG/Alb Conjugate
Zeta potential	−39.50 ± 3.54 A	−45.43 ± 2.06 B	−41.60 ± 1.05 AB	−45.30 ± 0.66 B	−43.83 ± 4.15 AB	−51.97 ± 0.81 C

Data represent mean ± standard deviation of three independent repeats. Different uppercase letters in each row indicate significant differences ($p < 0.05$).

3.8. Emulsion Properties

3.8.1. Emulsifying Activity Index (EAI) and Emulsion Stability (ES)

The ability of surfactants to form emulsion and to stabilize oil droplets after formation is quantified by EAI and ES, respectively [53,54]. An important techno-functional aspect of protein–polysaccharide conjugates is the improved interfacial properties induced by conjugation [55]. Generally, protein–polysaccharide conjugates have a higher ability than individual biopolymers or their mixtures to adsorb to the interface and to develop stable emulsions. Conjugation can also increase the solvation of hydrocolloids in the aqueous phase [56]. Emulsifying activity of conjugates depends on their ability to decrease the interfacial tension (IFT) between oil and water phases. IFT reduction is mediated by hydrophobic and hydrophilic moieties of conjugates toward oil and water, respectively. The EAI and turbidity of various protein/BSG mixtures and respective conjugates are shown in Figure 8. The EAIs of protein/BSG mixtures were significantly lower than those of their conjugated counterparts. In other words, the EAI in the mixture of WPI/BSG, SPI/BSG, and Alb/BSG increased from 12.94, 20.43, and 19.2 m^2/g to 79.32, 24.89, and 55.37 m^2/g, respectively. Polymerization during dry heat treatment plays an important role in the increase in emulsifying activity. Similarly to other emulsifiers, the adsorbed conjugates develop an interfacial film at the O/W interface, which is responsible for the emulsion stability. The type of protein had a significant effect on the ability of protein/BSG conjugates on emulsion stabilization. Emulsion samples prepared by different proteins/BSG mixtures and respective conjugates showed ES of 100% after 24 h of storage (Figure 9). After three days of storage, all proteins/BSG conjugates and the mixture of BSG with WPI revealed 100% stability. Water phase separation of around 27.7% and 38.3% was measured in the emulsion samples stabilized by SPI/BSG and Alb/BSG mixtures, respectively. Water phase separation at the bottom occurred in mixed and conjugated BSG-stabilized emulsions after 7 days of storage (Figure 9). The emulsion stabilized by the Alb/BSG conjugate revealed the lowest water phase separation. It was also homogenous, which indicated the Alb/BSG conjugate had the best emulsion stability properties. Lavaei et al. [57] studied the conjugation of gellan gum and soy protein by Maillard reaction. They reported that the conjugation increased the EAI and ES because of increasing protein solubility and improving the balance in the hydrophilic–hydrophobic character. Shen and Li [58] evaluated the functional properties of conjugated guar gum–pea protein isolate. The increase in the emulsion stability prepared by this conjugate was attributed to the formation of strong layer at the interface and thus improved steric stabilization.

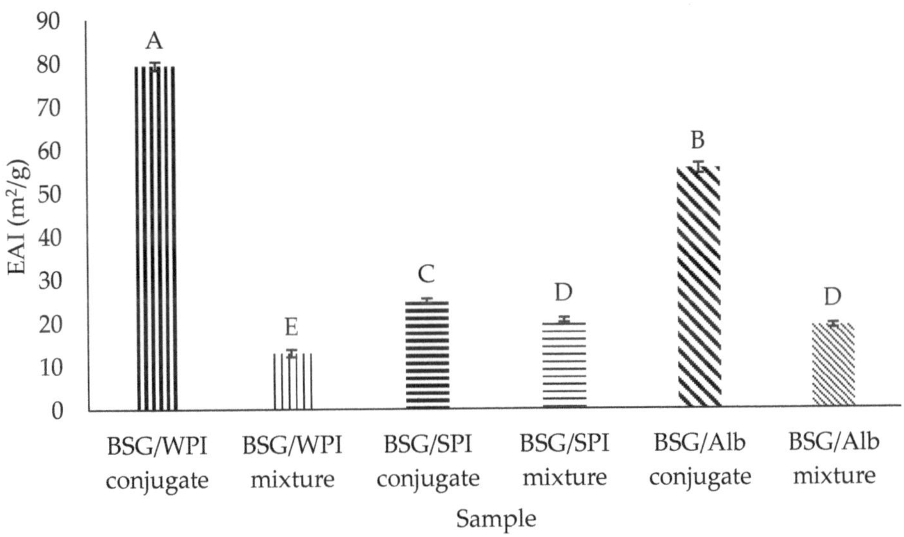

Figure 8. Emulsifying activity index of BSG/WPI, BSG/SPI, and BSG/Alb mixtures and respective conjugates. Different uppercase letters indicate significant differences ($p < 0.05$) between samples.

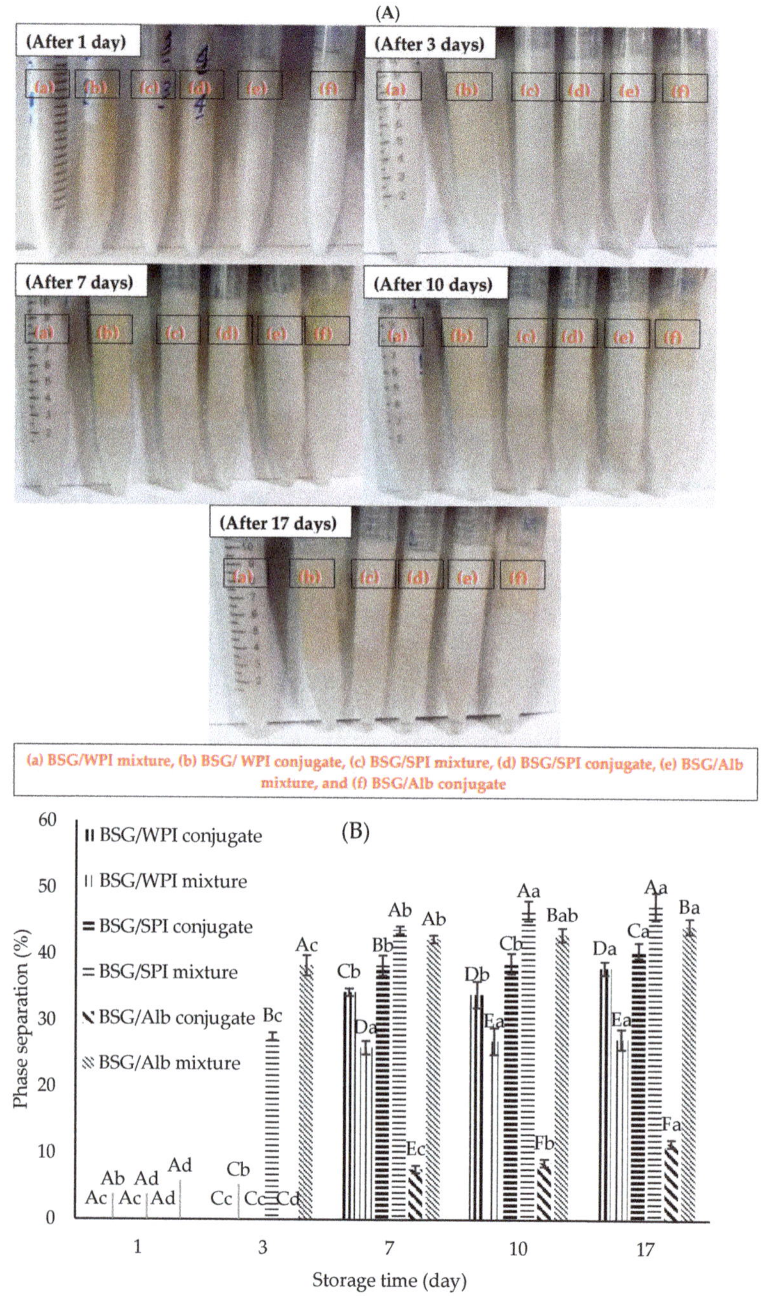

Figure 9. Visual observation of phase separation (**A**) and its quantification (**B**) in nanoemulsions stabilized by BSG/WPI, BSG/SPI, and BSG/Alb mixtures and conjugates. Different uppercase and lowercase letters indicate significant differences ($p < 0.05$) between samples at the same time and during the storage, respectively.

3.8.2. Droplet Size Distribution

Droplet size distribution plays an important role in the turbidity, stability, sensory experience, and rheology of food systems [59]. The volume-weighted mean droplet diameter of nanoemulsions and span (distribution width) values were measured over time and reported in Figures 10A–D and 11. Protein/BSG conjugation reduced the mean diameter and span values of nanoemulsions as compared to non-conjugated protein/polysaccharide mixture. This was attributed to the significantly increasing emulsifying activity after conjugation (Figure 4). The mean droplet diameter varied from 112.1 to 239.3 nm on day 3. Nanoemulsions prepared by the mixture of SPI/BSG and Alb/BSG exhibited the largest size at day 17, while those prepared by Alb/BSG conjugates showed the smallest size during storage (112.1 and 143.3 nm at day 3 and day 17, respectively). Droplet size reduction increases the emulsions stability against gravitational separation and some other phenomena such as coalescence and flocculation. As stated by Márquez et al. [60], the droplet size is related to the emulsifier's hydrophilic–lipophilic balance. These researchers reported that modified durian seed gum (DSG) prepared by the covalent linkage between DSG and WPI had an improved hydrophilic–hydrophobic character compared to native DSG. Akhtar and Dickinson [17] found that the conjugates of maltodextrin/WPI had high emulsifying activity and were able to prepare emulsions with smaller droplet size. Introducing hydrophobic moieties to conjugates might facilitate the adsorption rate at the interface and also lead to decreased droplet coalescence and aggregation [61,62]. The droplet size of nanoemulsions stabilized by the mixture of SPI/BSG and Alb/BSG significantly increased during storage; however, it remained relatively constant in nanoemulsions prepared by the WPI/BSG mixture and all conjugated protein/BSG counterparts. The results of span values did not show any significant difference between all samples during the storage. The span values of nanoemulsions ranged from 0.1 to 0.27 at day 3 and 0.15 to 0.32 at day 17 (Figure 11). The lower values indicated a narrower size distribution (more homogeneity).

3.8.3. Zeta Potential

Table 5 reports the zeta potential values of nanoemulsions stabilized by different proteins/BSG mixtures and/or conjugates. In all samples, the O/W interface was negatively charged, arising from the negative charge along BSG backbone as well as the net negative charge of proteins at pH values above the isoelectric point. Generally, the absolute values of zeta potential decreased during storage; however, the final values were still above the minimum required values (i.e., $|\text{zeta potential}| > 20$ mV) for providing sufficient electrostatic repulsion among the oil droplets. Considering the high values of the zeta potential and the physical instability of nanoemulsions during storage, it can be concluded that additional factors (e.g., adsorption affinity, viscosity, interfacial film characteristics, etc.) other than the electrostatic repulsion among the dispersed droplets (i.e., zeta potential) are required for the long term stability of nanoemulsions. Chen et al. [63] evaluated the conjugation of pectin and albumin using the Maillard reaction to increase the emulsifying properties and reported that the conjugates had lower zeta potential values than the respective electrostatic complex. These researchers concluded that the involvement of more amino acid residues for the covalent linkage with carboxyl groups of pectin in the conjugates is responsible for lower values of zeta potential. Our finding is not in agreement with that of Chen, Ji, Qiu, Liu, Zhu, and Yin [63]. Except for nanoemulsions stabilized by the mixture and the conjugate of BSG/Alb, other nanoemulsions stabilized by the mixture or conjugate of same biopolymers did not significantly differ in zeta potential values after preparation. The conjugation reaction is expected to occur between the amino groups of proteins (i.e., present in the side chain of amino acid residues such as ε-amino group of lysine and/or terminal amino group of proteins) and the reducing end (i.e., carbonyl group) of polysaccharides and not carboxyl groups along backbone. Moreover, other factors such as pH, ionic strength, and microviscosity have a significant contribution to electrophoretic mobility. Similarly to our finding in the mixture and conjugate of BSG/Alb, Feng et al. [64] reported that the emulsions stabilized by the conjugates of SPI-GA had higher zeta potential absolute values than those stabilized by a non-conjugated mixture. This might be due to more blocking proteins' amino groups after conjugation (Table 4) and hence modulating the net charge of conjugated biopolymers (i.e., the increase in contribution of negative charges). In our study, the highest physical stability was observed in the emulsions stabilized by the BSG/Alb conjugate.

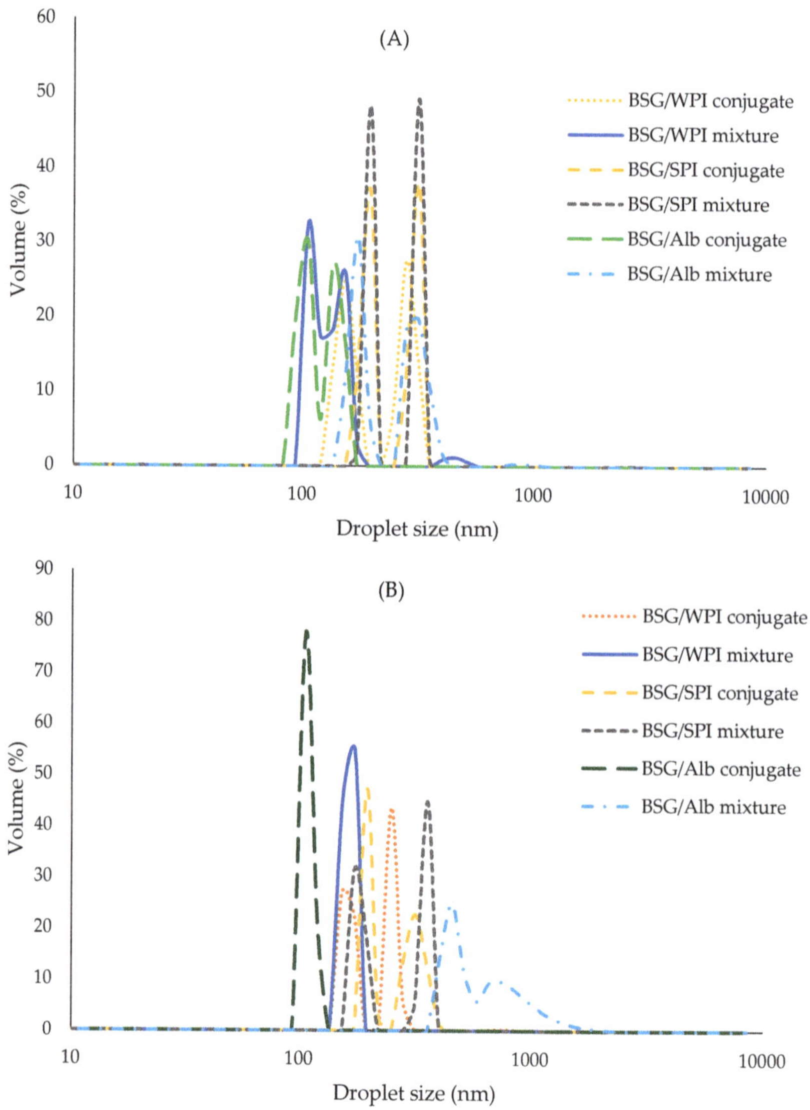

Figure 10. *Cont.*

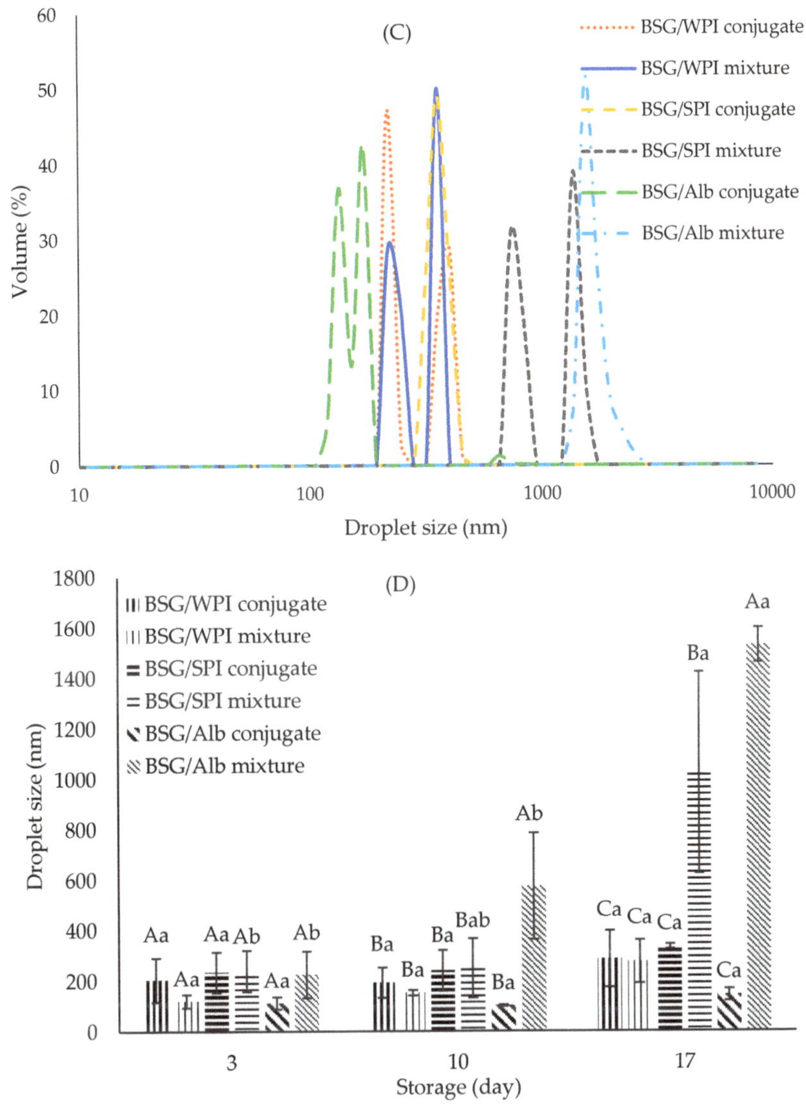

Figure 10. (**A–C**): droplet size distribution of nanoemulsions stabilized by the mixtures and/or conjugates of BSG/WPI, BSG/SPI, and BSG/Alb, respectively, during storage for 3, 10, and 17 days; (**D**) reports mean droplet diameter (nm) during storage. Different uppercase and lowercase letters indicate significant differences ($p < 0.05$) between samples at the same time and during the storage, respectively.

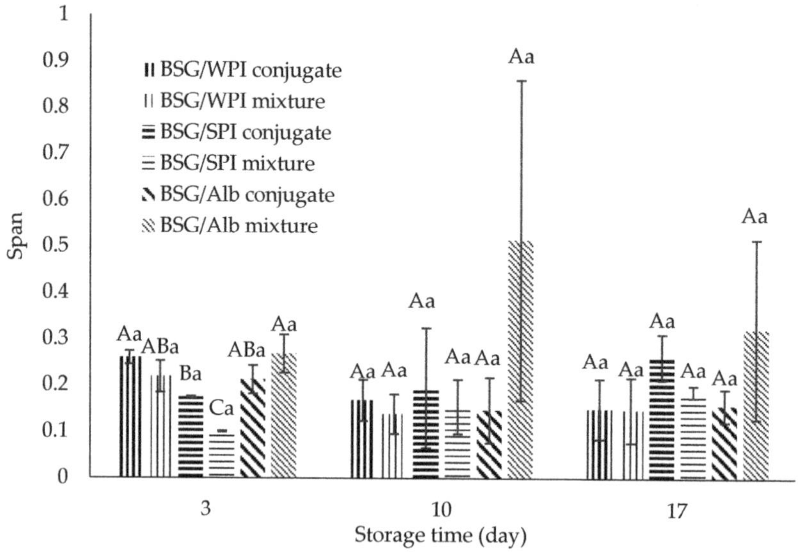

Figure 11. Span values of nanoemulsions stabilized by the mixtures and/or conjugates of BSG/WPI, BSG/SPI, and BSG/Alb during storage for 3, 10, and 17 days. Different uppercase and lowercase letters indicate significant differences ($p < 0.05$) between samples at the same time and during the storage, respectively.

Table 5. Zeta potential (mV) values of nanoemulsions stabilized by the mixtures and/or conjugates of BSG/WPI, BSG/SPI, and BSG/Alb during storage for 3, 7, 10, and 17 days.

Sample	Storage Time (day)			
	3	7	10	17
BSG/WPI conjugate	−61.50 ± 0.61 Ab	−58.53 ± 2.47 Bab	−57.70 ± 0.14 BCa	−57.60 ± 1.37 CDa
BSG/WPI mixture	−60.23 ± 1.61 Aa	−58.33 ± 2.64 Ba	−58.40 ± 1.51 BCa	−57.83 ± 1.48 CDa
BSG/SPI conjugate	−61.70 ± 1.35 Ab	−57.77 ± 1.60 ABa	−56.13 ± 0.81 Ba	−56.17 ± 2.01 BCa
BSG/SPI mixture	−60.60 ± 1.45 Ab	−54.23 ± 2.47 Aa	−53.67 ± 1.00 Aa	−54.13 ± 0.68 ABa
BSG/Alb conjugate	−73.23 ± 2.97 Bc	−67.13 ± 2.25 Cb	−59.77 ± 2.03 Ca	−58.90 ± 1.57 Da
BSG/Alb mixture	−61.37 ± 1.80 Ab	−54.77 ± 0.84 ABa	−53.77 ± 0.49 Aa	−53.13 ± 0.61 Aa

Data are the average of three independent replicates ± standard deviation. Different uppercase letters in each column and lowercase ones in each row indicate significant differences ($p < 0.05$).

3.9. Foaming Properties

Figure 12 shows the effect of conjugation on the foaming capacity of samples. The foaming capacity of WPI (78.57%), SPI (61.91%), and Alb (71.43%) in their mixtures with BSG increased to 107.14%, 85.71%, and 85.71%, respectively, after making conjugates with BSG. Therefore, conjugation increased the foaming capacity. Conjugation increases protein solubility and stability with consequent benefits for emulsification characteristics. The higher foaming capacity of conjugates compared to mixtures might be due to higher solubility, and better action as a surfactant at the air–water interface in the presence of the emulsifying and thickening effects of grafted BSG. Hamdani et al. [65] evaluated the effect of conjugation on functional properties of guar gum using egg white lysozyme. They reported that the conjugation of lysozyme improved foaming capacity (12%) and foam stability (35%). Koshani, Aminlari, Niakosari, Farahnaky, and Mesbahi [11] reported that the conjugation of tragacanthin and lysozyme enhanced foaming capacity (52.25%) and foam stability (58.23%). Also, Hashemi, Aminlari, and Moosavinasab [12] reported that the conjugation of lysozyme and xanthan gum improved foaming capacity (49.90%) and foam stability (37.80%).

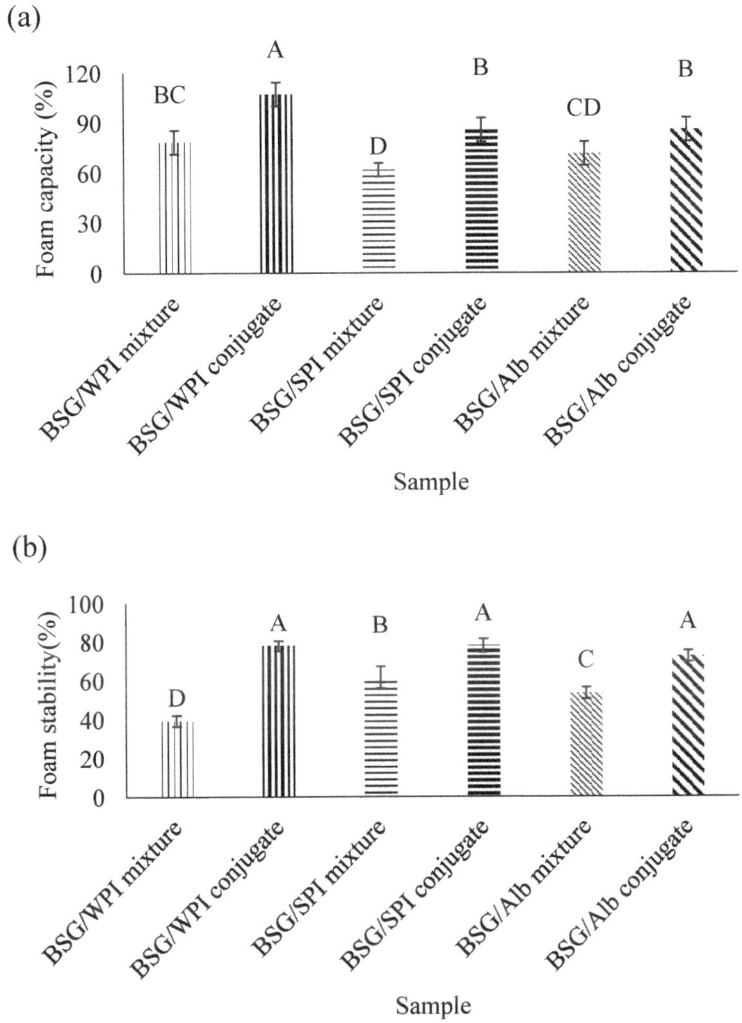

Figure 12. Foaming capacity (**a**) and foam stability (**b**) of BSG/WPI mixture, BSG/WPI conjugate, BSG/SPI mixture, BSG/SPI conjugate, BSG/Alb mixture, and BSG/Alb conjugate. Different uppercase letters indicate significant differences ($p < 0.05$) between samples.

Foam stability is determined as the ability of foam to maintain some of its properties during storage [66]. The gas bubbles rise to the top and may undergo deformation to form polyhedral structures [67]. In this study, the foam stability was significantly improved by conjugation with BSG (Figure 12). The foam stability of WPI (39.34%), SPI (61.57%), and Alb (53.37%) in their mixtures with BSG (non-conjugated condition) increased to 77.86%, 77.91%, and 72.32%, respectively, after formation of conjugates with BSG. The reason for better foam stabilization by BSG/protein conjugates compared to the BSG/protein mixture is likely that it improves the viscoelastic properties of interfacial film at the air–water interface in the presence of BSG. Non-grafted BSG also plays a role through increasing the viscosity of water phase. Previous studies reported that the viscosity has a significant effect on the foam stability [68]. Biopolymer networks increase the viscosity of bulk-phase and reduce the coalescence of air bubbles which increases the foam stability [67].

4. Conclusions

In the present research, the conjugates of WPI, SPI, and Alb with BSG were fabricated using the dry heating method via a Maillard-type reaction. The conjugates showed higher emulsifying activity and emulsion stability than non-conjugated binary mixtures of biopolymers. The BSG-albumin conjugate was the best sample. Conjugation also improved the emulsion properties. The conjugates exhibited tremendous functional properties with better foaming activities than non-conjugated samples. Furthermore, the presence of basil seed gum prevented some adverse effects of heating on proteins in powder form. Modification of natural gums can be applied for improving the techno-functional properties in food systems.

Supplementary Materials: The following supporting information can be downloaded at: https://www.mdpi.com/article/10.3390/foods14030390/s1, Figure S1. Scanning electron micrographs of (a) BSG/WPI mixture, (b) BSG/WPI conjugate, (c) BSG/SPI mixture, (d) BSG/SPI conjugate, (e) BSG/Alb mixture, and (f) BSG/Alb conjugate. Magnification 500×; scale bar = 100 μm.

Author Contributions: Conceptualization, H.H., S.M.H.H. and M.H.E.; methodology, H.H., S.M.H.H. and M.H.E.; software, H.H., S.M.H.H. and M.H.E.; validation, H.H., S.M.H.H., M.H.E., M.K., M.-T.G. and M.N.; formal analysis, H.H., S.M.H.H., M.H.E., M.K., M.-T.G. and M.N.; investigation, H.H., S.M.H.H., M.H.E., M.K., M.-T.G. and M.N.; resources, H.H., S.M.H.H., M.H.E., M.K., M.-T.G. and M.N.; data curation, H.H., S.M.H.H., M.H.E., M.K., M.-T.G. and M.N.; writing—original draft preparation, H.H., S.M.H.H., M.H.E., M.K., M.-T.G. and M.N.; writing—review and editing, H.H., S.M.H.H., M.H.E., M.K., M.-T.G. and M.N.; visualization, H.H., S.M.H.H., M.H.E., M.K., M.-T.G. and M.N.; supervision, S.M.H.H. and M.H.E.; project administration, H.H., S.M.H.H., M.H.E., M.K., M.-T.G. and M.N.; funding acquisition, S.M.H.H. and M.H.E. All authors have read and agreed to the published version of the manuscript.

Funding: This research was funded by Shiraz University (Grant number 96GCU5M194065).

Institutional Review Board Statement: Not applicable.

Informed Consent Statement: Not applicable.

Data Availability Statement: The original contributions presented in this study are included in the article/Supplementary Materials. Further inquiries can be directed to the corresponding author.

Acknowledgments: This study was supported by Shiraz University.

Conflicts of Interest: The authors declare no conflicts of interest.

References

1. Malakar, M.; Mandal, D. *Ocimum* spp.(basil): An incredible plant. In *Advances in Medicinal and Aromatic Plants*; Apple Academic Press: Cambridge, MA, USA, 2024; pp. vol1: 89–vol81: 164.
2. Calderón Bravo, H.; Vera Céspedes, N.; Zura-Bravo, L.; Muñoz, L.A. Basil seeds as a novel food, source of nutrients and functional ingredients with beneficial properties: A review. *Foods* **2021**, *10*, 1467. [CrossRef] [PubMed]
3. Naji-Tabasi, S.; Razavi, S.M.A.; Mohebbi, M.; Malaekeh-Nikouei, B. New studies on basil (*Ocimum bacilicum* L.) seed gum: Part i–fractionation, physicochemical and surface activity characterization. *Food Hydrocoll.* **2016**, *52*, 350–358. [CrossRef]
4. Hosseini-Parvar, S.; Matia-Merino, L.; Goh, K.; Razavi, S.M.A.; Mortazavi, S.A. Steady shear flow behavior of gum extracted from *Ocimum basilicum* L. Seed: Effect of concentration and temperature. *J. Food Eng.* **2010**, *101*, 236–243. [CrossRef]
5. Naji-Tabasi, S.; Razavi, S.M.A. New studies on basil (*Ocimum bacilicum* L.) seed gum: Part ii—Emulsifying and foaming characterization. *Carbohydr. Polym.* **2016**, *149*, 140–150. [CrossRef]
6. Liu, Y.; Zhao, G.; Zhao, M.; Ren, J.; Yang, B. Improvement of functional properties of peanut protein isolate by conjugation with dextran through maillard reaction. *Food Chem.* **2012**, *131*, 901–906. [CrossRef]
7. Gumus, C.E.; Davidov-Pardo, G.; McClements, D.J. Lutein-enriched emulsion-based delivery systems: Impact of maillard conjugation on physicochemical stability and gastrointestinal fate. *Food Hydrocoll.* **2016**, *60*, 38–49. [CrossRef]
8. Cai, B.; Saito, A.; Ikeda, S. Maillard conjugation of sodium alginate to whey protein for enhanced resistance to surfactant-induced competitive displacement from air–water interfaces. *J. Agric. Food Chem.* **2018**, *66*, 704–710. [CrossRef]

9. Bi, B.; Yang, H.; Fang, Y.; Nishinari, K.; Phillips, G.O. Characterization and emulsifying properties of β-lactoglobulin-gum acacia seyal conjugates prepared via the maillard reaction. *Food Chem.* **2017**, *214*, 614–621. [CrossRef]
10. Kim, D.-Y.; Shin, W.-S. Functional improvements in bovine serum albumin–fucoidan conjugate through the maillard reaction. *Food Chem.* **2016**, *190*, 974–981. [CrossRef]
11. Koshani, R.; Aminlari, M.; Niakosari, M.; Farahnaky, A.; Mesbahi, G. Production and properties of tragacanthin-conjugated lysozyme as a new multifunctional biopolymer. *Food Hydrocoll.* **2015**, *47*, 69–78. [CrossRef]
12. Hashemi, M.M.; Aminlari, M.; Moosavinasab, M. Preparation of and studies on the functional properties and bactericidal activity of the lysozyme–xanthan gum conjugate. *LWT-Food Sci. Technol.* **2014**, *57*, 594–602. [CrossRef]
13. Zhao, M.; He, H.; Ma, A.; Hou, T. Sources, chemical synthesis, functional improvement and applications of food-derived protein/peptide-saccharide covalent conjugates: A review. *Crit. Rev. Food Sci. Nutr.* **2023**, *63*, 5985–6004. [CrossRef] [PubMed]
14. Aziznia, S.; Askari, G.; Emamdjomeh, Z.; Salami, M. Effect of ultrasonic assisted grafting on the structural and functional properties of mung bean protein isolate conjugated with maltodextrin through maillard reaction. *Int. J. Biol. Macromol.* **2024**, *254*, 127616. [CrossRef]
15. Ali, N.; Saeidy, S. Maillard modification. In *Physicochemical and Enzymatic Modification of Gums: Synthesis, Characterization and Application*; Springer: Berlin/Heidelberg, Germany, 2022; pp. 77–98.
16. Hashemi, M.M.; Aminlari, M.; Forouzan, M.M.; Moghimi, E.; Tavana, M.; Shekarforoush, S.; Mohammadifar, M.A. Production and application of lysozyme-gum arabic conjugate in mayonnaise as a natural preservative and emulsifier. *Pol. J. Food Nutr. Sci.* **2018**, *68*, 33–43. [CrossRef]
17. Akhtar, M.; Dickinson, E. Whey protein–maltodextrin conjugates as emulsifying agents: An alternative to gum arabic. *Food Hydrocoll.* **2007**, *21*, 607–616. [CrossRef]
18. Zhang, S.; Wang, K.; Qin, Y.; Zhu, S.; Gao, Q.; Liu, D. The synthesis, biological activities and applications of protein–polysaccharide conjugates in food system: A review. *Food Qual. Saf.* **2023**, *7*, fyad006. [CrossRef]
19. Siddiquy, M.; JiaoJiao, Y.; Rahman, M.H.; Iqbal, M.W.; Al-Maqtari, Q.A.; Easdani, M.; Yiasmin, M.N.; Ashraf, W.; Hussain, A.; Zhang, L. Advances of protein functionalities through conjugation of protein and polysaccharide. *Food Bioprocess Technol.* **2024**, *17*, 2077–2097. [CrossRef]
20. Zhang, X.; Chen, M.; Wang, N.; Luo, J.; Li, M.; Li, S.; Hemar, Y. Conjugation of chitopentaose with β-lactoglobulin using maillard reaction, and its effect on the allergic desensitization in vivo. *Int. J. Biol. Macromol.* **2024**, *258*, 128913. [CrossRef]
21. Urango, A.C.M.; Meireles, M.A.A.; Silva, E.K. Maillard conjugates produced from proteins and prebiotic dietary fibers: Technological properties, health benefits and challenges. *Trends Food Sci. Technol.* **2024**, *147*, 104438. [CrossRef]
22. Zhang, X.; Wang, Y.; Li, Z.; Li, Y.; Qi, B. Effects of polysaccharide type on the structure, interface behavior, and foam properties of soybean protein isolate hydrolysate-polysaccharide maillard conjugates. *Food Hydrocoll.* **2024**, *151*, 109801. [CrossRef]
23. Tao, L.; Wang, H.; Wang, J.; Zhang, J.; Yu, L.; Song, S. Characterization and emulsion stability of soybean protein isolate/soybean peptide and ginseng polysaccharide conjugates. *LWT* **2024**, *196*, 115860. [CrossRef]
24. Amiratashani, F.; Yarmand, M.S.; Kiani, H.; Askari, G.; Naeini, K.K.; Parandi, E. Comprehensive structural and functional characterization of a new protein-polysaccharide conjugate between grass pea protein (*Lathyrus sativus*) and xanthan gum produced by wet heating. *Int. J. Biol. Macromol.* **2024**, *254*, 127283. [CrossRef] [PubMed]
25. Kan, X.; Hu, Y.; Huang, Y.; Fan, X.; Chen, G.; Ye, H.; Zeng, X. Characterization of whey protein isolate-gum Arabic Maillard conjugate and evaluation of the effects of conjugate-stabilized emulsion on microbiota of human fecal cultures. *Food Hydrocoll.* **2023**, *134*, 108060. [CrossRef]
26. Laemmli, U.K. Cleavage of structural proteins during the assembly of the head of bacteriophage t4. *Nature* **1970**, *227*, 680. [CrossRef]
27. Chen, B.; Chen, L.; Li, C.; Huang, W.; Zhao, Y.; Ai, C.; Teng, H. Ultrasound-assisted glycosylation of ovalbumin and dextran conjugate carrier for anthocyanins and their stability evaluation. *Ultrason. Sonochem.* **2024**, *109*, 107024. [CrossRef]
28. Gahruie, H.H.; Eskandari, M.H.; Khalesi, M.; Van der Meeren, P.; Hosseini, S.M.H. Rheological and interfacial properties of basil seed gum modified with octenyl succinic anhydride. *Food Hydrocoll.* **2020**, *101*, 105489. [CrossRef]
29. Kheynoor, N.; Hosseini, S.M.H.; Yousefi, G.-H.; Gahruie, H.H.; Mesbahi, G.-R. Encapsulation of vitamin c in a rebaudioside-sweetened model beverage using water in oil in water double emulsions. *LWT* **2018**, *96*, 419–425. [CrossRef]
30. Gahruie, H.H.; Ziaee, E.; Eskandari, M.H.; Hosseini, S.M.H. Characterization of basil seed gum-based edible films incorporated with zataria multiflora essential oil nanoemulsion. *Carbohydr. Polym.* **2017**, *166*, 93–103. [CrossRef]
31. Hosseini, S.M.H.; Gahruie, H.H.; Razmjooie, M.; Sepeidnameh, M.; Rastehmanfard, M.; Tatar, M.; Naghibalhossaini, F.; Van der Meeren, P. Effects of novel and conventional thermal treatments on the physicochemical properties of iron-loaded double emulsions. *Food Chem.* **2019**, *270*, 70–77. [CrossRef]
32. Wang, L.; Cheng, X.; Zhang, S.; Dongye, Z.; Kang, M.; Li, Z.; Chen, C.; Qian, Y.; Ren, Y. The rheological/interfacial behavior and stability properties of nanoemulsions prepared using whey protein-carboxymethyl chitosan conjugates. *Colloids Surf. A Physicochem. Eng. Asp.* **2023**, *662*, 130924. [CrossRef]

33. Al-Lafi, A.G.; Isam, A.-N. Application of 2d-cos-ftir spectroscopic analysis to milk powder adulteration: Detection of melamine. *J. Food Compos. Anal.* **2022**, *113*, 104720. [CrossRef]
34. Colthup, N. *Introduction to Infrared and Raman Spectroscopy*; Elsevier: Amsterdam, The Netherlands, 2012.
35. Moremedi, T.; Katata-Seru, L.; Sardar, S.; Bandyopadhyay, A.; Makhado, E.; Hato, M.J. Application of synthetic monomers grafted xanthan gum for rhodamine b removal in aqueous solution. *Int. J. Mater. Metall. Eng.* **2020**, *14*, 123–131.
36. Tayel, A.A.; Ebaid, A.M.; Otian, A.M.; Mahrous, H.; El Rabey, H.A.; Salem, M.F. Application of edible nanocomposites from chitosan/fenugreek seed mucilage/selenium nanoparticles for protecting lemon from green mold. *Int. J. Biol. Macromol.* **2024**, *273*, 133109. [CrossRef]
37. Mohammadi, A.; Ghorbani, M.; Sadeghi Mahoonak, A.; Jafari, S.M. Improving the oxidative stability of purslane seed oil via emulsions stabilized by whey protein isolate-inulin mixtures and conjugates. *Innov. Food Technol.* **2024**, *11*, 226–248.
38. Silverstein, R.M.; Webster, F.X.; Kiemle, D.J.; Bryce, D.L. *Spectrometric Identification of Organic Compounds*; John Wiley & Sons: Hoboken, NJ, USA, 2014.
39. Zhang, Q.; Zhou, Y.; Yue, W.; Qin, W.; Dong, H.; Vasanthan, T. Nanostructures of protein-polysaccharide complexes or conjugates for encapsulation of bioactive compounds. *Trends Food Sci. Technol.* **2021**, *109*, 169–196. [CrossRef]
40. Niu, L.Y.; Jiang, S.T.; Pan, L.J.; Zhai, Y.S. Characteristics and functional properties of wheat germ protein glycated with saccharides through maillard reaction. *Int. J. Food Sci. Technol.* **2011**, *46*, 2197–2203. [CrossRef]
41. Yadav, M.P.; Strahan, G.D.; Mukhopadhyay, S.; Hotchkiss, A.T.; Hicks, K.B. Formation of corn fiber gum–milk protein conjugates and their molecular characterization. *Food Hydrocoll.* **2012**, *26*, 326–333. [CrossRef]
42. Pirestani, S.; Nasirpour, A.; Keramat, J.; Desobry, S.; Jasniewski, J. Structural properties of canola protein isolate-gum arabic maillard conjugate in an aqueous model system. *Food Hydrocoll.* **2018**, *79*, 228–234. [CrossRef]
43. Diftis, N.; Kiosseoglou, V. Improvement of emulsifying properties of soybean protein isolate by conjugation with carboxymethyl cellulose. *Food Chem.* **2003**, *81*, 1–6. [CrossRef]
44. Nakamura, S.; Gohya, Y.; Losso, J.N.; Nakai, S.; Kato, A. Protective effect of lysozyme-galactomannan or lysozyme-palmitic acid conjugates against *Edwardsiella tarda* infection in carp, *Cyprinus carpio* L. *FEBS Lett.* **1996**, *383*, 251–254. [CrossRef]
45. Kasran, M.; Cui, S.W.; Goff, H.D. Covalent attachment of fenugreek gum to soy whey protein isolate through natural maillard reaction for improved emulsion stability. *Food Hydrocoll.* **2013**, *30*, 552–558. [CrossRef]
46. Johnson, C.M. Differential scanning calorimetry as a tool for protein folding and stability. *Arch. Biochem. Biophys.* **2013**, *531*, 100–109. [CrossRef] [PubMed]
47. Ibanoglu, E.; Erçelebi, E.A. Thermal denaturation and functional properties of egg proteins in the presence of hydrocolloid gums. *Food Chem.* **2007**, *101*, 626–633. [CrossRef]
48. Boye, J.I.; Alli, I.; Ismail, A.A. Interactions involved in the gelation of bovine serum albumin. *J. Agric. Food Chem.* **1996**, *44*, 996–1004. [CrossRef]
49. Takahashi, K.; Lou, X.-F.; Ishii, Y.; Hattori, M. Lysozyme-glucose stearic acid monoester conjugate formed through the maillard reaction as an antibacterial emulsifier. *J. Agric. Food Chem.* **2000**, *48*, 2044–2049. [CrossRef]
50. Gahruie, H.H.; Mirzapour, A.; Ghiasi, F.; Eskandari, M.H.; Moosavi-Nasab, M.; Hosseini, S.M.H. Development and characterization of gelatin and persian gum composite edible films through complex coacervation. *LWT* **2022**, *153*, 112422. [CrossRef]
51. Vardhanabhuti, B.; Ikeda, S. Isolation and characterization of hydrocolloids from monoi (*Cissampelos pareira*) leaves. *Food Hydrocoll.* **2006**, *20*, 885–891. [CrossRef]
52. Sadar, L.N. Rheological and Textural Characteristics of Copolymerized Hydrocolloidal Solutions Containing Curdlan Gum. Master's Thesis, University of Maryland, College Park, MD, USA, 2004.
53. Liang, R.-H.; Wang, L.-H.; Chen, J.; Liu, W.; Liu, C.-M. Alkylated pectin: Synthesis, characterization, viscosity and emulsifying properties. *Food Hydrocoll.* **2015**, *50*, 65–73. [CrossRef]
54. Wang, W.-D.; Li, C.; Chen, C.; Fu, X.; Liu, R.H. Effect of chitosan oligosaccharide glycosylation on the emulsifying property of lactoferrin. *Int. J. Biol. Macromol.* **2022**, *209*, 93–106. [CrossRef]
55. Kan, X.; Chen, G.; Zhou, W.; Zeng, X. Application of protein-polysaccharide maillard conjugates as emulsifiers: Source, preparation and functional properties. *Food Res. Int.* **2021**, *150*, 110740. [CrossRef]
56. Nooshkam, M.; Varidi, M.; Zareie, Z.; Alkobeisi, F. Behavior of protein-polysaccharide conjugate-stabilized food emulsions under various destabilization conditions. *Food Chem. X* **2023**, *18*, 100725. [CrossRef] [PubMed]
57. Lavaei, Y.; Varidi, M.; Nooshkam, M. Gellan gum conjugation with soy protein via maillard-driven molecular interactions and subsequent clustering lead to conjugates with tuned technological functionality. *Food Chem. X* **2022**, *15*, 100408. [CrossRef] [PubMed]
58. Shen, Y.; Li, Y. Acylation modification and/or guar gum conjugation enhanced functional properties of pea protein isolate. *Food Hydrocoll.* **2021**, *117*, 106686. [CrossRef]
59. Ghayour, N.; Hosseini, S.M.H.; Eskandari, M.H.; Esteghlal, S.; Nekoei, A.-R.; Gahruie, H.H.; Tatar, M.; Naghibalhossaini, F. Nanoencapsulation of quercetin and curcumin in casein-based delivery systems. *Food Hydrocoll.* **2019**, *87*, 394–403. [CrossRef]

60. Márquez, A.L.; Palazolo, G.G.; Wagner, J.R. Water in oil (w/o) and double (w/o/w) emulsions prepared with spans: Microstructure, stability, and rheology. *Colloid Polym. Sci.* **2007**, *285*, 1119–1128. [CrossRef]
61. Wang, H.; Williams, P.A.; Senan, C. Synthesis, characterization and emulsification properties of dodecenyl succinic anhydride derivatives of gum arabic. *Food Hydrocoll.* **2014**, *37*, 143–148. [CrossRef]
62. Niakousari, M.; Damyeh, M.S.; Gahruie, H.H.; Bekhit, A.E.D.A.; Greiner, R.; Roohinejad, S. Conventional emulsions. In *Emulsion-Based Systems for Delivery of Food Active Compounds: Formation, Application, Health and Safety*; Wiley: Hoboken, NJ, USA, 2018; pp. 1–27. [CrossRef]
63. Chen, H.; Ji, A.; Qiu, S.; Liu, Y.; Zhu, Q.; Yin, L. Covalent conjugation of bovine serum album and sugar beet pectin through maillard reaction/laccase catalysis to improve the emulsifying properties. *Food Hydrocoll.* **2018**, *76*, 173–183. [CrossRef]
64. Feng, S.; Guo, Y.; Liu, F.; Li, Z.; Chen, K.; Handa, A.; Zhang, Y. The impacts of complexation and glycated conjugation on the performance of soy protein isolate-gum Arabic composites at the o/w interface for emulsion-based delivery systems. *Food Hydrocoll.* **2023**, *135*, 108168. [CrossRef]
65. Hamdani, A.M.; Wani, I.A.; Bhat, N.A.; Siddiqi, R.A. Effect of guar gum conjugation on functional, antioxidant and antimicrobial activity of egg white lysozyme. *Food Chem.* **2018**, *240*, 1201–1209. [CrossRef]
66. Naji, S.; Razavi, S.M.; Karazhiyan, H. Effect of thermal treatments on functional properties of cress seed (*Lepidium sativum*) and xanthan gums: A comparative study. *Food Hydrocoll.* **2012**, *28*, 75–81. [CrossRef]
67. Jahanbin, K.; Moini, S.; Gohari, A.R.; Emam-Djomeh, Z.; Masi, P. Isolation, purification and characterization of a new gum from *Acanthophyllum bracteatum* roots. *Food Hydrocoll.* **2012**, *27*, 14–21. [CrossRef]
68. Koocheki, A.; Taherian, A.R.; Bostan, A. Studies on the steady shear flow behavior and functional properties of *Lepidium perfoliatum* seed gum. *Food Res. Int.* **2013**, *50*, 446–456. [CrossRef]

foods

Article

Anti-Diabetic Effect of Soy–Whey Dual-Protein on Mice with Type 2 Diabetes Mellitus Through INS/IRS1/PI3K Signaling Pathway

Na Li, Hu Li, Duo Feng, Mengjie Li, Di Han, Tianxin Liu and Jing Wang *

Institute of Food and Nutrition Development, Ministry of Agriculture and Rural Affairs, Beijing 100081, China; 15910807606@163.com (N.L.); lihu01@caas.cn (H.L.); 15525926785@163.com (D.F.); limj0804@126.com (M.L.); handi87@126.com (D.H.); liutianxin2022@163.com (T.L.)
* Correspondence: wangjing07@caas.cn

Abstract: The effects of soy protein and whey protein supplementation on glycemic control show inconsistency, and the mechanisms underlying the impact of a high-protein diet on blood glucose regulation remain unclear. This study aimed to explore the impact of a dual-protein (DP) blend comprising soy protein isolate (SPI) and whey protein concentrate (WPC), processed through high-pressure homogenization, on mice with Type 2 diabetes mellitus (T2DM) and its potential mechanisms. In the in vitro experiments, an insulin-resistant (IR) HepG2 cell model was treated with DP, resulting in a significant enhancement of glucose uptake and upregulation of IRS1 and GLUT4 expression. For the in vivo experiments, male C57BL/6J mice were randomly assigned into four groups (n = 6) based on body weight: normal control, T2DM model group, Metformin-treated group, and DP-treated group. Following a 5-week feeding period, Metformin and DP significantly reduced levels of blood sugar, AUC, TC, TG, and LDL-C in T2DM mice. Additionally, TP and ALB levels in the DP group were notably higher in the model group. In the liver and pancreas, DP alleviated histopathological changes and promoted liver glycogen synthesis in T2DM mice. Moreover, the levels of IRS1 and PI3K in the livers of mice in the DP group were significantly higher than those in the model group. Compared with the model groups, DP significantly reduced the expression of CD45 and increased the expression of CD206 in the pancreas of mice. Furthermore, 16S rRNA analysis revealed that DP altered the composition of the gut microbiota in diabetic mice, increasing the relative abundance of *Lactobacillus*, *Parvibacter*, and *Lactobacillaceae*. This suggested that DP could alleviate functional metabolic disorders in the gut and potentially reverse the risk of related complications. In conclusion, soy whey dual-protein may have an effective nutritional therapeutic effect on T2DM mice by regulating lipid metabolism, the INS/IRS1/PI3K signaling pathway, and gut microbiota.

Keywords: dual-protein; type 2 diabetes; insulin resistance; mechanism of action; gut microbiota

1. Introduction

Type 2 diabetes mellitus (T2DM) is a metabolic disorder syndrome characterized by hyperglycemia resulting from defective insulin secretion, which can induce chronic damage and functional impairment in tissues such as blood vessels, kidneys, and the nervous system [1]. Currently, oral hypoglycemic medications, including metformin, sulfonylureas, and α-glucosidase inhibitors, remain effective methods for controlling the progression of T2DM [2,3]. However, long-term use of these drugs by T2DM patients can lead to numerous adverse effects, such as gastrointestinal intolerance or metabolic acidosis [4]. Notably, dietary and nutritional interventions have proven to be effective in managing the condition of T2DM patients and improving their quality of life, representing one of the emerging and promising therapeutic approaches with minimal side effects for the treatment of T2DM [5].

Protein supplements or high-protein diets have been proven to effectively improve blood sugar regulation [6]. Dietary proteins and amino acids influence insulin sensitivity and regulate glucose transport and utilization in the body through a series of complex biochemical reactions [7]. Studies have shown that whey protein could reduce risk factors for T2DM by enhancing satiety, promoting insulin secretion to lower postprandial blood glucose levels, and facilitating weight loss, thereby preventing obesity and the onset of T2DM [8]. However, the specific mechanisms remain unclear. Additionally, soy protein and peptides could act on various molecular targets, such as α-glucosidase, α-amylase, and dipeptidyl peptidase IV, as well as intracellular signaling pathways in multiple organs, to maintain glucose homeostasis and improve impaired glucose and lipid metabolism [9]. Although high-quality proteins like soy protein and whey protein have long been widely used in nutritional interventions, only a limited number of studies have reported their effects on metabolic diseases, with inconsistent conclusions and mechanisms that still require further exploration [10–15]. Dual-protein foods are a new dietary concept proposed in recent years with the aim of supplementing high-quality protein. It is of great significance to promote the combination of protein resources from different sources by using high-tech method means to improve their nutritional quality [16,17]. High-pressure homogenization (HPH) is widely used in the food industry to produce emulsified foods, such as soft drinks, milk, and condiments. During the high-pressure homogenization process, protein molecules are exposed to cavitation, shearing, turbulence, and heating in a short period of time, which can disrupt protein aggregates and change protein structure, thereby affecting their solubility and other functional properties [18]. Our previous research found that HPH could improve the functional properties of proteins by altering the structure of soy–whey dual-proteins [19]. However, whether the alteration of protein function will have an intervention effect on T2DM remains to be verified.

The gut microbiota plays a crucial role in the dynamic ecosystem of the human body, and its composition can be influenced by numerous factors, such as the host and the environment. Numerous studies have confirmed that imbalances in gut microbiota are not only closely associated with intestinal diseases but also significantly linked to metabolic disorders such as diabetes [20,21]. Therefore, modulating the gut microbial community is considered an important aspect of preventing or slowing down the progression of T2DM. Studies have shown that high-protein diets or protein-based nutritional interventions could influence the composition and abundance of gut microbiota. This is primarily attributed to the proteolytic fermentation of undigested proteins by bacterial proteases in the colon. Subsequently, metabolites released by the gut microbiota can affect host physiology [22]. Our previous research had demonstrated that DP supplementation increases the diversity of gut microbiota and alters microbial abundance patterns in patients undergoing allogeneic hematopoietic stem cell transplantation. It could also alleviate osteoporosis by modulating the gut microbiota, indicating its potential for regulating gut microbial communities [23,24]. Therefore, we hypothesized that DP nutritional intervention could reverse T2DM by modulating the gut microbiota.

Herein, this study investigated the effects of DP on alleviating insulin resistance (IR) and T2DM both in vitro and in vivo, and explored the role of gut microbiota in improving metabolic health under DP intervention. We induced insulin resistance in HepG2 cells

(IR-HepG2) and tested the efficacy of DP in ameliorating cellular insulin resistance. Subsequently, a T2DM mouse model was established using a high-fat diet (HFD) combined with streptozotocin (STZ) to investigate the effects of DP on improving glucose, lipid, and protein metabolism. Additionally, 16S rDNA amplicon sequencing was employed to further explore the composition of gut microbiota and key regulatory factors underlying the improvement of metabolic health by DP.

2. Materials and Methods

2.1. Chemicals and Reagents

SUPRO 787 IP SPI (protein content, 90%) was obtained from IFF Inc. (Shanghai, China). WPC (protein content, 80%) was obtained from Arla Foods (DANEMARK).

Insulin was purchased from Shanghai Aladdin Biochemical Technology Co., Ltd. (Shanghai, China) and metformin from Haohong Biopharmaceutical Technology Co., Ltd. (Shanghai, China). 3-(4,5-dimethylthiazol-2-yl)-2,5-diphenyltetrazolium bromide (MTT) and sodium dodecyl sulfate (SDS) were purchased from Beyotime (Shanghai, China). Glucose kit (glucose oxidase method) was purchased from Nanjing Jiancheng Biotechnology Co., Ltd. (Nanjing, China).

STZ and metformin was purchased from Sigma-Aldrich (St. Louis, MO, USA). IRS1 Rabbit mAb, PI3K Rabbit mA, and GLUT4 Rabbit mA were purchased from Cell Signaling Technology(Boston, US). Goat Anti-Rabbit IgG H&L (Alexa Fluor® 488) (ab150077) and Goat Anti-Rabbit IgG H&L (Alexa Fluor® 555) (ab150078) were purchased from Abcam(Cambridge, Britain). 2-(4-amidinophenyl)-1H-indole-6-carboxamidine (DAPI) was purchased from Solarbio, Beijing, China. PE anti-mouse CD11b antibody, FITC anti-mouse CD45 antibody, PE anti-mouse CD3 antibody, FITC anti-mouse CD8 antibody, and APC anti-mouse CD4 antibody were purchased from Biolegend, USA. All other reagents were analytical reagents from chemical companies in China.

2.2. Soy–Whey Dual-Protein Solution Sample Preparation

Our team previously conducted a study on the application of high-pressure homogenization to dual-protein [19]. On this basis, the DP optimization treatment is carried out. Soy protein isolate (SPI) and whey protein concentrate (WPC) were weighed, respectively, and according to the team's preliminary results, they passed through the planetary ball mill at a ratio of 1:1. The rotational speed was set at 500 rpm/min, the intermittent time was 30 min, and the mixed protein powder was obtained by grinding and mixing for 8 h. Then, the DP powder was weighed and then dissolved in deionized water to prepare a 6% (*w*/*w*) DP solution. The solution was stirred magnetically for 1.5 h. Then, the container was sealed with plastic wrap and allowed to hydrate in a refrigerator at 4 °C for 10 h. After standing and returning to room temperature, the filtrate was homogenized under 60 MPa high pressure through 60 mesh sieves, and each pressure cycle was three times. After homogenization, the pH of the solution was adjusted to 7 to obtain the final DP sample. The amino acid profile of DP is detailed in Supplementary Table S1.

2.3. Cell Experiments

2.3.1. Cell Culture

Human hepatocellular carcinoma cell line HepG2 cells were acquired from Shanghai Cell Bank of the Chinese Academy of Sciences. The cell lines were cultured in DMEM, which contained 10% fetal bovine serum (FBS; Hyclone, Thermo Scientific, Waltham, MA, USA), penicillin (100 U/mL), and streptomycin (100 μg/mL) (Gran Island, NY, USA). The incubator conditions were set at 37 °C and 5% CO_2. A 0.25% trypsin was used for digestion and passage.

2.3.2. Cytotoxicity

A total of 5 × 10^3 cells/well were seeded on 96 well plates and allowed to adhere overnight. The cells were treated with increasing concentrations of the sample diluted in cell media achieving a total volume of 100 μL. The cells were incubated with Met and increasing concentrations of DP at 37 °C for 24 h. Next, the medium with serum was replaced with fresh medium containing (3-(4,5-dimethylthiazol-2-yl)-2,5-diphenyltetrazolium bromide) (MTT, 10 μL of a 5 mg/mL solution in PBS buffer) and the cells were further incubated for 4 h. 10% SDS solution was then added (100 μL/well) and the plates were kept in the dark for an additional 12 h. Absorption measurements were performed on a Bio-Rad plate reader at 570 nm (peak absorbance) and at 650 nm (background absorbance).

2.3.3. Glucose Uptake

HepG2 cells were seeded in 96-well plates in DMEM supplemented with 10% (*v*/*v*) FBS. The density of the cells in each plate was 5 × 10^3 cells/well. Preliminary experiments were conducted to establish insulin resistance model in HepG2 cells in accordance with Zheng et al. [25]. We removed the original medium when the cells became adherent and replaced it with DMEM containing 10^{-6} mol/L insulin. The model of insulin resistance (IR-HepG2) was established after 24 h. Control group, insulin resistance model group (IR), metformin (Met) group (0.03 mg/mL), and DP group (1.00, 5.00, 10.00 mg/mL) were set up in the experiment, with three parallel cells in each group. The culture was continued for 24 h, and then the glucose content in the culture medium supernatant was determined by glucose detection kit.

2.3.4. In Vitro Fluorescence Staining

HepG2 cells were seeded on 24-well plates (1 × 10^4 cells/well) and allowed to adhere overnight. The IR-HepG2 model was established after 24h induction with 10^{-6} mol/L insulin. The cells were then treated with DP (10 mg/mL) and metformin (0.03 mg/mL) for 24 h. PBS served as control treatment. The above cells were washed with PBS and further incubated with IRS1 Rabbit Monoclonal Antibody for 2 h at 37 °C. Subsequently, the cells were incubated with Goat Anti-Rabbit IgG H&L (Alexa Fluor® 488) for 1 h at 37 °C. The expression of IRS1 inside the cells was observed by CLSM.

2.3.5. Western Blot

HepG2 cells were seeded in 6-well plates (1 × 10^6 cells/well) and allowed to adhere overnight. The IR-HepG2 cell model was established after 24 h induction with 10^{-6} mol/L insulin. The cells were treated with DP (10 mg/mL) and metformin (0.03 mg/mL) for 24 h. After washing the cells with cold PBS three times, RIPA lysis buffer containing protease and phosphatase inhibitors was added to each well. The proteins of cells were extracted through centrifuge at a speed of 12,000 rpm for 5 min. Protein content quantification was carried out by the BCA protein assay kit. Then, the electrophoreses process was conducted through SDS-PAGE by a gel-electrophoretic apparatus (Bio-Rad mini, Hercules, CA, USA), and the proteins were transferred to the PVDF films and incubated with the antibodies against various proteins overnight on a shaker at 4 °C. Subsequently, the PVDF films were washed five times and incubated with HRP conjugated antibodies for 1 h. The western blot images were obtained by Amersham Imager 600 (AI600, General Electric Co., Ltd., Boston, MA, USA) with 300 μL of ECL chemiluminescent reagent (Beyotime Biotechnology Co., Ltd., Shanghai, China) added on the top of the membrane.

2.4. Animal Experiments

Six-week-old male C57BL/6J mice (approximately 16 to 18 g) were purchased from Sibeifu (Beijing) Biotechnology Co., Ltd. (Beijing, China). Mice were housed no more than six per cage under standard laboratory conditions (12 h light/dark cycle at room temperature [approximately 22–24 °C]) with free access to clean water and a standard diet. The entirety of animal experiments conducted received approval from the Animal Management and Ethics Committee of the Institutional Animal Care and Use Committee, Institute of Food and Nutrition Development, Ministry of Agriculture and Rural Affairs, as per the stipulated license number 20240315-05.

The whole animal experimental design is summarized in Figure 1. All C57BL/6J mice were adaptively administered with common chow for a week and further replaced with a high-fat diet (HFD) for 6 weeks. The ingredients of HFD were 66.5% common diet, 20% sucrose, 10% fat from lard, 2.5% cholesterol, and 1% sodium deoxycholate (SPF (Beijing, China) BIOTECHNOLOGY Co., Ltd. (Fuxin, China)).

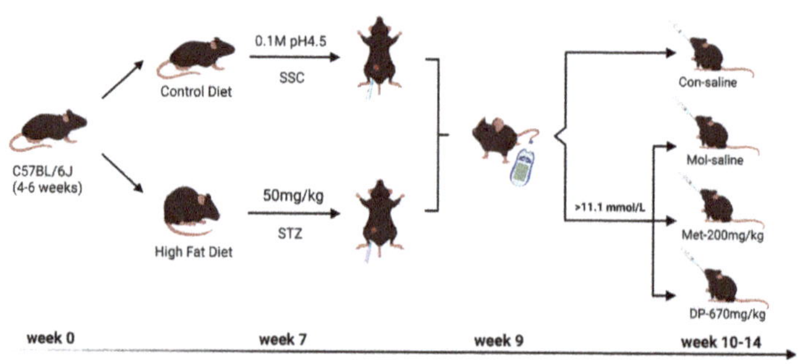

Figure 1. Animal experimental design. Con: normal control mice; Mol: diabetic model mice; Met: diabetic mice treated with metformin (200 mg/kg); DP: diabetic mice treated with DP (670 mg/kg). Figure created with Biorender.com.

Streptozotocin (STZ) was intraperitoneally injected into mice on days 1 and 3 of the 7th week at a dose of 50 mg/kg. Mice were required to fast for 12 h before STZ injection. STZ was dissolved in Citrate Buffered Saline (SSC) (pH 4.2–4.5). Notably, it required temporary preparation for use and protection from light and was placed in an ice bath during the injection. Fasting blood glucose (FBG) level was measured in the 9th week, and mice with FBG > 11.1 mmol/L were considered to be successfully modeled with diabetes. Furthermore, they were randomly divided into a T2DM model (Mol) group, metformin (Met) group, and dual-protein (DP) group based on the FBG level, and C57BL/6J mice reared on common chow for the same period comprised the normal control (Con) group (6 mice/group). Mice in the Met group were given metformin orally (200 mg/kg/day), and mice in DP group were given DP solution orally (670 mg/kg/day). Equal amounts of saline were administered by oral gavage to both the Con and Mol groups mice for 5 weeks consecutively.

2.4.1. Body Weight, FBG, and Glucose Tolerance Test (OGTT)

Body weight and FBG were measured every week during the experimental period. In the last week of the experimental period, the animals were subjected to an oral glucose tolerance test. In brief, mice were fasted for 12 h and administered an oral glucose load (2 g/kg of body weight), and then the blood glucose level was measured using a glucometer (Sinocare) at 0, 30, 60, 90, and 120 min, respectively. The areas under the curve (AUCs) of blood glucose levels were calculated to evaluate glucose tolerance.

2.4.2. Analysis of Serum Biochemistry

Total cholesterol (TC), triacylglycerols (TG), high-density lipoprotein cholesterol (HDL-C), low-density lipoprotein cholesterol (LDL-C), total protein (TP), and albumin (ALB) were examined by an Automatic biochemical analyzer (Chemray 800, Shenzhen, China).

2.4.3. Detection of Immunohistochemistry and Immunofluorescence

Livers and pancreas were collected and a slice of each one was fixated with 4% paraformaldehyde. Liver and pancreas were stained with hematoxylin and eosin (H&E) and the liver was stained with Periodic-acid Schiff (PAS) staining. The histological images were taken using a CLSM (LSM-800, ZEISS, Oberkochen, Germany).

For immunofluorescence analysis, frozen sections of pancreas were incubated for 14 h at 4 °C with a primary antibody for insulin (ab7842, Abcam), glucagon (ab10988, Abcam), and then with secondary antibodies at room temperature for 2 h. Slides were observed under a laser scanning confocal microscope. The area of cells positive for insulin antibody in the islets were counted and used to calculate the positive cell ratio.

Immunohistochemistry (IHC) was used to analyze the expression of IRS1 and PI3K in the mouse livers. Specifically, the wax-embedded livers were cut into 5 μm sections. We used anti-mouse IRS1 (Servicebio, Wuhan, China) and PI3K (Servicebio, Wuhan, China) antibody (1:100) at 4 °C overnight for IHC staining. All slices were counterstained with hematoxylin at 25 °C for 1 min. microscopic images were obtained with a light microscope. Three random areas from each sample were chosen and analyzed using Image J (National Institutes of Health, Bethesda, MD, USA).

2.4.4. Detection of CD45 in Pancreas Slices by CLSM

Following the previously described treatment, pancreas slices from each group of animals were collected. The tissue was incubated with CD45+-specific fluorescent probe. The presence of CD45+ T cells were assessed using CLSM (LSM-800, ZEISS, Germany).

2.4.5. Detection of CD45 T Cells in Pancreas by Flow Cytometry

Fresh pancreatic tissues were collected for antidiabetic immune response analysis via FACS. Briefly, samples were dissociated into single-cell suspensions, and then red blood cells were removed with red blood cell lysing buffer (Beyotime). After that, samples were blocked with 0.1% BSA in PBS followed by incubation with relevant antibodies for 1 h at room temperature. For characterizing T cells and leukocyte in pancreas, cells were stained by anti-CD3-PE, anti-CD4-PC5.5, anti-CD8-FITC, and anti-CD45-FITC. All the antibodies used above were all purchased from Bio Legend. Flow cytometric data acquisition was performed with CytExpert-2.6 software, and the data were processed using FlowJo-v10 software. Data were expressed as mean ± SD (n = 3).

2.4.6. 16S rDNA Sequencing Analysis

In this experiment, Illumina novaseq6000 sequencing platform was used to amplify and sequence the V3-V4 variable region. First, the mouse feces DNA was extracted by the kit, and the purity and concentration were detected. After the V3–V4 variable region was selected, PCR amplification was performed using a specific primer with Barcode and a high-fidelity DNA polymerase. PCR products were detected by 2% agarose-gel electrophoresis, and target fragments were recovered by using the Quant-iT PicoGreen dsDNA Assay Kit. Then, the recovered products were detected and quantified by FLx800 fluorescence quantification system. The TruSeq Nano DNA LT Library Prep Kit was used to construct the library. In order to ensure that the library quality met the standards of on-machine sequencing, Agilent Bioanalyzer 2100 and Promega QuantiFluor were used for quality inspection, and after passing the inspection, on-machine sequencing was performed.

Raw data obtained by sequencing has a certain proportion of dirty data. In order to make the results of information analysis more accurate and reliable, quality control analysis such as quality filtering, noise reduction, splicing, and chimera removal were performed on the raw data by using Qiime2 default parameters, and sequences with abundance less than 10 are filtered out (all samples are summed up), so as to obtain ASVs. Based on the flattened ASVs, ASV diversity index analysis and sequencing depth detection could be performed. Based on taxonomic information, statistical analysis of community structure could be performed at various taxonomic levels.

Sequence analysis was carried out using UPARSE software package and UPARSE-OTU and UPARSE-OTUref algorithms. Sequences with similarity ≥97% were assigned to the same OTU. We selected representative sequences for each OTU and label each representative sequence with classification information using the RDP classifier.

2.5. Statistical Analysis

Data were expressed as mean ± SD. All the data were analyzed using GraphPad Prism 9.0 (GraphPad Software, Inc. La Jolla, CA, USA). Statistical significance was analyzed using a one-way analysis of variance (ANOVA) followed by the Bonferroni test. Data were considered statistically significant for $p < 0.05$.

3. Result

3.1. Effects of DP on the Insulin Resistance (IR)-HepG2 Cell Model

HepG2 cells are commonly utilized for investigating hepatic insulin signaling pathways and glucose homeostasis in vitro. The cytotoxicity of DP on HepG2 cells was evaluated using the MTT assay after 24 h to establish its effective concentration (Figure 2A). Experimental groups were exposed to varying concentrations of DP (5, 10, 15, and 20 mg/mL), while a Con group remained DP-free. The findings demonstrated a biphasic effect of DP on cell growth and differentiation, with promotion at lower concentrations and inhibition at higher concentrations. Specifically, DP at 5 mg/mL enhanced HepG2 cell growth and differentiation, yielding a cell viability rate of 105.98% compared to the Con group. Conversely, at 10 mg/mL, DP initiated inhibition of HepG2 cell growth, resulting in a cell viability rate of 91.45%. These outcomes suggest that DP concentrations below 10 mg/mL have negligible impact on HepG2 cell viability.

To examine the impact of DP on glucose uptake and cell viability in insulin-induced HepG2 cells (IR-HepG2 model), we quantified glucose levels in the culture supernatant using a glucose oxidase assay kit and evaluated cell viability (Figure 2B). Our findings revealed that glucose uptake was 6.95 mmol/L in the Con group, whereas in the IR group treated with 10^{-6} M insulin, glucose uptake significantly decreased to 3.52 mmol/L ($p < 0.01$), confirming the successful establishment of the IR-HepG2 cell model. DP at concentrations of 1, 5, and 10 mg/mL significantly enhanced glucose uptake in IR-HepG2 cells compared to the IR group, with higher concentrations yielding greater improvements. Notably, at a DP concentration of 10 mg/mL, glucose uptake reached 6.43 mmol/L, surpassing the effect of the positive control metformin (0.03 mg/mL), albeit without statistical significance. These results indicate the development of insulin resistance in the IR group, while highlighting the potential of DP at specific concentrations to ameliorate glucose utilization in insulin-resistant HepG2 cells in a dose-dependent manner.

To further explore the mechanism of DP on HepG2 cells, we utilized laser scanning confocal microscopy (CLSM) to assess the expression of IRS1 in the IR-HepG2 cell model and conducted semi-quantitative analysis (Figure 2C,D). Our CLSM findings demonstrated a reduction in IRS1 protein levels in HepG2 cells to 57.30% of the Con group under insulin stimulation. After intervention with metformin (0.03 mg/mL) and DP (10 mg/mL), the expression levels of IRS1 protein were restored by 44.74% and 51.4%, respectively. These results indicate that DP enhances glucose uptake by elevating IRS1 expression, thereby significantly enhancing glucose metabolism. Furthermore, to evaluate the translocation of GLUT4 protein in the IR-HepG2 cell model, we assessed GLUT4 protein expression through western blotting (Figure 2E). All western blot data were normalized against GAPDH as a housekeeping protein control. Band intensities were quantified using ImageJ(x64) software (version 1.8.0_60) and presented as ratios of target protein to GAPDH (Figure 2F). Our data indicated a significant reduction in GLUT4 expression in the Mol group post insulin induction compared to the Con group. However, following a 24-h treatment with DP (10 mg/mL), GLUT4/GAPDH increased by 1.62-fold relative to Mol group ($p < 0.01$). These findings suggest that DP facilitates GLUT4 translocation in the IR-HepG2 cell model, leading to enhanced glucose transport into the cells and mitigation of insulin resistance.

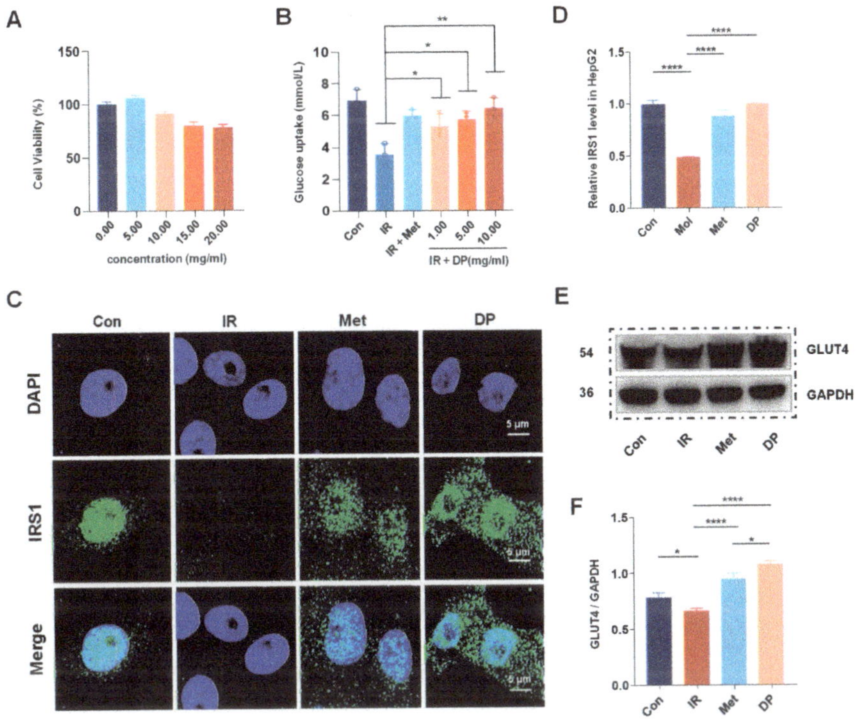

Figure 2. In vitro DP promotes glucose uptake and activates IRS1/GLUT4 pathway in HepG2 cells. (**A**) Cell viability of HepG2 treated with DP in 24 h. (**B**) Glucose intake in different treatment groups in 24 h. (**C**) Representative CSLM images of IRS1 expression on HepG2 treated with metformin (0.03 mg/mL) and DP (10 mg/mL). The cell nuclei were stained with DAPI (blue) and IRS1 were stained with Alexa Fluor® 488 (green). (**D**) Relative IRS1 level in HepG2. (**E**) Western blot analysis to detect the expression levels of GLUT4 treated by metformin (0.03 mg/mL) and DP (10 mg/mL) for 24 h, and GAPDH was used as an internal reference protein. (**F**) Quantitative western blot analysis of GLUT4 by Image J software (n = 3, data are presented as mean ± SD). Statistical significances between every two groups were calculated via one-way ANOVA. * $p < 0.05$, ** $p < 0.01$, **** $p < 0.0001$.

3.2. *Effects of DP on Blood Glucose, Body Weight, and Food Intake in T2DM Mice*

To investigate the impact of DP on blood glucose levels in T2DM mice, fasting blood glucose (FBG) levels were measured weekly for five consecutive weeks (Figure 3A). Compared to the normal mice in the Con group, the FBG levels in T2DM mice were significantly elevated ($p < 0.01$) and exceeded 11.1 mmol/L, indicating the successful establishment of a stable T2DM mouse model induced by HFD/STZ. After five weeks of intervention with DP and metformin, the FBG levels in the Met group decreased more significantly ($p < 0.01$), dropping from 11.57 mmol/L to 6.78 mmol/L. Similarly, the FBG levels in the DP intervention group also showed a significant reduction ($p < 0.01$), decreasing from 11.52 mmol/L to 8.13 mmol/L.

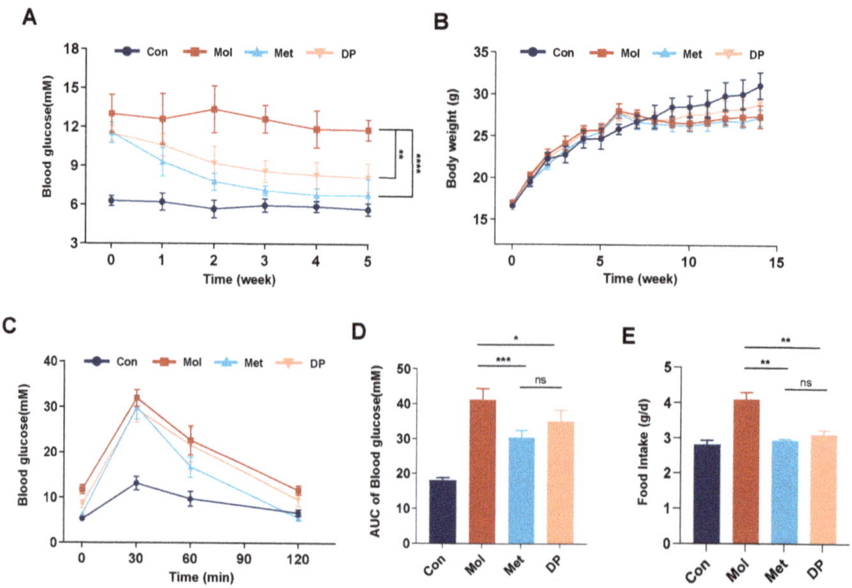

Figure 3. Effect of DP on glucose metabolism in T2DM mice. The weekly changes of blood glucose (**A**) and body weight (**B**) in T2DM mice during DP intervention. (**C**) OGTT test performed at the end of the experiment. All mice were fasted overnight, and then orally administered a 2.0 g/kg dose of glucose. The blood glucose levels were measured at 0, 30, 60, 90, and 120 min. (**D**) The area under the curve (AUC) of each group. (**E**) Food intake of mice after intervention (n = 6, data are presented as mean ± SD). Statistical significances between every two groups were calculated via one-way ANOVA. * $p < 0.05$, ** $p < 0.01$, *** $p < 0.001$, **** $p < 0.0001$.

Changes in body weight are shown in Figure 3B. Following six weeks of HFD feeding, the average body weight of the Mol group increased from 16.93 g to 27.98 g. Subsequent to the induction of the T2DM model, the mice displayed symptoms including polydipsia, polyphagia, polyuria, and weight loss. Their FBG levels surpassed 11.1 mmol/L, with random blood glucose levels exceeding 16.7 mmol/L, thereby confirming the successful establishment of the T2DM model. Weekly monitoring of body weight indicated that, in comparison to the Con group, the Mol group mice exhibited a consistent decline in weight. Following treatment with metformin and DP, their body weight commenced recovery and displayed a continuous increase. Of note, the body weight of the DP group mice was higher than that of the model group, suggesting that DP could ameliorate weight loss in mice with T2DM. Conversely, the weight gain rate in the Met group was lower than that in the other groups, possibly attributable to the gastrointestinal side effects associated with metformin.

The oral glucose tolerance test (OGTT) was utilized to assess fluctuations in blood glucose levels in mice over a 2 h period following the oral administration of a high dose of glucose (Figure 3C,D). The OGTT, along with the calculation of the area under the curve (AUC), serves as indicators of the glucose tolerance of the mice. To explore the impact of DP on abnormal glucose tolerance in HFD/STZ-induced T2DM mice, an OGTT was conducted on four mouse groups following five weeks of continuous intervention. Results from the OGTT indicated a noteworthy distinction in the rate of glucose level elevation within 30 min post-administration of a 20% glucose solution between the Con group and Mol group. The AUC was computed, revealing a significant reduction in the AUC with DP intervention compared to the Mol group ($p < 0.05$). Moreover, the Met group exhibited an even more substantial decrease in AUC ($p < 0.01$). These findings suggest that DP has the potential to ameliorate abnormal OGTT and AUC in T2DM mice, thereby enhancing their glucose tolerance. Furthermore, as depicted in Figure 3E, there was a notable decrease in food intake among T2DM mice ($p < 0.01$), indicating that both metformin and DP could effectively mitigate the heightened food consumption observed in T2DM mice.

3.3. Effects of DP on Lipid and Protein Metabolism in Mice with T2DM

Diabetes commonly co-occurs with disruptions in lipid and protein metabolism, with lipid metabolism intricately linked to insulin and glucose regulation in the body. Figure 4A–D illustrate the concentrations of TC, TG, HDL-C, and LDL-C. Our findings indicate that TC, TG, and LDL-C levels were markedly elevated in mice with T2DM compared to those in normal mice. Conversely, treatment with DP significantly decreased TC ($p < 0.01$), TG ($p < 0.01$), and LDL-C levels ($p < 0.05$), while no significant difference was observed in HDL-C levels.

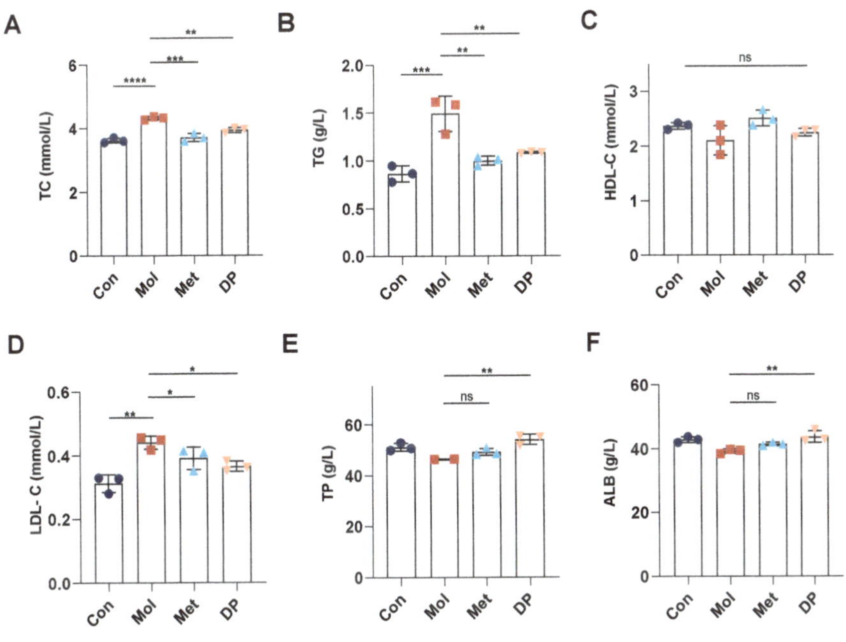

Figure 4. Effect of DP on lipid metabolism and protein metabolism in mice with T2DM. TC (**A**), TG (**B**), HDL-C (**C**), LDL-C (**D**), TP (**E**), and ALB (**F**) concentration in serum after intervention (n = 3, data are presented as mean ± SD). Statistical significances between every two groups were calculated via one-way ANOVA. * $p < 0.05$, ** $p < 0.01$, *** $p < 0.001$, **** $p < 0.0001$, ns—no significant difference.

In Figure 4E,F, protein metabolism indicators were assessed. In T2DM mice, compared to the Con group, serum levels of TP and ALB significantly decreased to 46.34 g/L and 39.22 g/L, respectively ($p < 0.05$). Following metformin intervention, TP concentration increased to 49.26 g/L, showing no significant difference compared to the Con and Mol groups. In the DP group, TP concentration rose to 54.03 g/L, significantly higher than in the Mol group ($p < 0.01$) and slightly higher than the control group's 51.18 g/L. ALB levels in the metformin group increased to 41.46 g/L, with no significant difference compared to the model group. Conversely, ALB concentration in the DP group increased to 43.71 g/L, showing a significant difference compared to the model group ($p < 0.01$).

3.4. Effects of DP on Liver Injury in Mice with T2DM

Hematoxylin and eosin (H&E) staining of liver tissue sections was conducted to assess the hepatoprotective effects of DP (Figure 5A). In healthy mice, liver histology displayed a standard lobular structure with hepatocytes encircling the central vein. Conversely, T2DM mice exhibited pronounced fat accumulation and extensive lipid vacuolation in the liver. The results suggest that DP supplementation markedly ameliorated histopathological alterations and microvesicular steatosis in the liver.

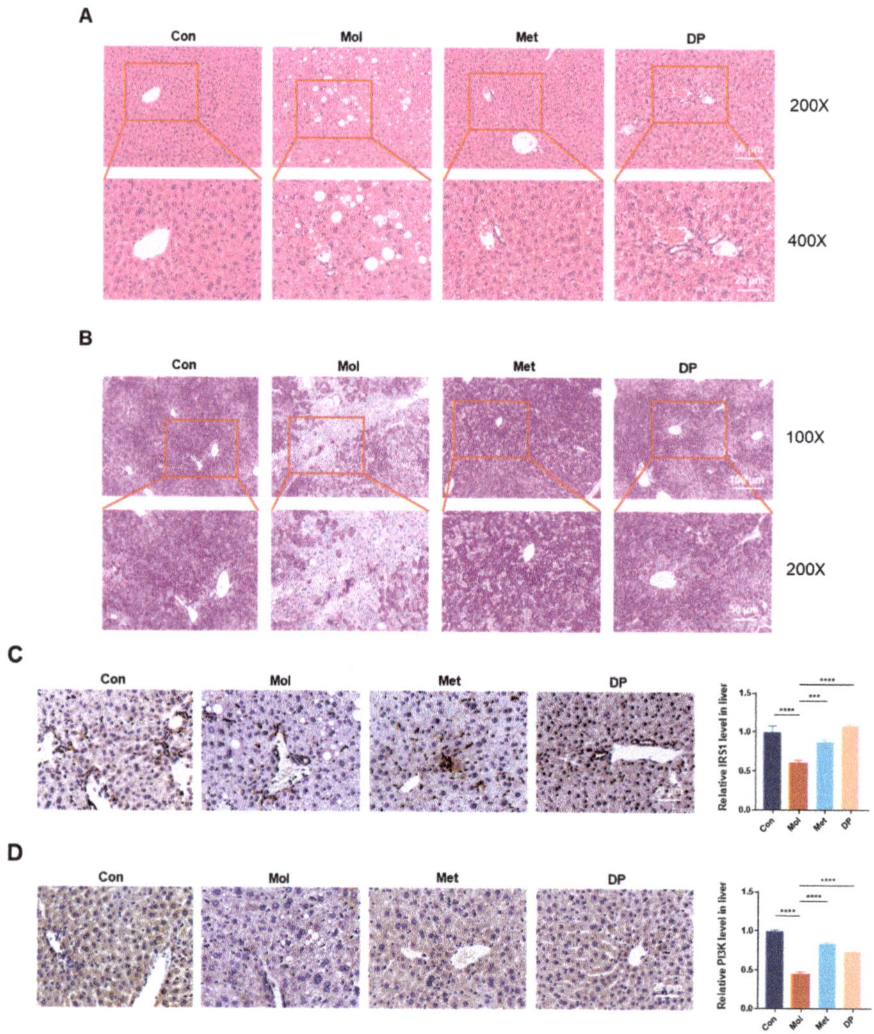

Figure 5. Effect of DP on liver histopathology and the relative expression of IRS1 and PI3K on liver. (**A**) HE staining study of liver tissues in each treatment group. (**B**) PAS staining of liver tissue of mice in each treatment group. Scale bar = 100 μm. (**C**) The relative expression of IRS1 on liver. Quantitative

analysis was conducted using Image J software. (**D**) The relative expression of PI3K on liver. Scale bar = 20 μm. Quantitative analysis was conducted using Image J software (n = 3, data are presented as mean ± SD). Statistical significances between every two groups were calculated via one-way ANOVA. *** $p < 0.001$, **** $p < 0.0001$.

Periodic acid–Schiff (PAS) staining was utilized to assess the impact of DP on hepatic glycogen levels in the liver (Figure 5B). Following PAS staining, hepatic glycogen was observed in various shades of purple, while cell nuclei were stained blue. Liver tissue samples from the control group exhibited a consistent deep purple hue, indicating high glycogen levels in healthy mice. In contrast, liver tissue staining in the Mol group displayed a notably lighter shade. Following DP intervention, the intensity of glycogen staining in liver tissues increased, with the most pronounced staining observed in the metformin-treated group. These findings suggest that both metformin and DP supplementation can enhance hepatic glycogen levels, potentially by facilitating glycogen synthesis and modulating hepatic glycogen metabolism in T2DM mice, thereby contributing to improved glycemic control.

To further investigate the mechanism of DP intervention, we examined its impact on the expression of IRS1 (Figure 5C) and PI3K (Figure 5D) proteins through immunohistochemical staining, followed by semi-quantitative analysis using ImageJ software. In comparison to the Con group, the levels of IRS1 and PI3K proteins in the livers of Mol group mice were significantly decreased ($p < 0.05$), indicating impaired insulin receptor pathways and insulin signaling transduction in mice with T2DM. Following DP intervention, the expression levels of IRS1 and PI3K proteins were restored by 1.77-fold and 1.6-fold respectively, compared to the Mol group. These findings suggest that DP may mitigate IR by activating the IRS1/PI3K signaling pathway.

3.5. Effects of DP on Pancreatic Tissue in Mice with T2DM

Islet cell damage is a key feature of T2DM, and morphological changes in islets can be observed through H&E staining of pancreatic tissue. Figure 6A showed that in normal mice, islet cells exhibited a regular shape, clear boundaries, and an intact structure. Conversely, islets from T2DM mice exhibit notable changes, such as islet cell atrophy and morphological variations. A semi-quantitative analysis of islet cell area using ImageJ software indicates that both metformin and DP interventions effectively mitigate islet damage in T2DM mice to a comparable degree. There is a significant increase in the islet area compared to the Mol group ($p < 0.01$), with a trend towards improved islet structure integrity. These outcomes suggest that DP has the potential to ameliorate islet damage in T2DM.

In addition to causing islet cell damage, T2DM disrupts the regulation of insulin secretion, leading to insufficient insulin production [26]. Islet cells were stained using fluorescence-labeled insulin and glucagon antibodies, with nuclei counterstained using DAPI. As shown in Figure 6B, insulin (green fluorescence) was primarily localized in the central region of the islets, secreted by β-cells. Conversely, glucagon (red fluorescence) was mainly located in the peripheral region, secreted by α-cells. The results demonstrated that both metformin and DP interventions significantly promoted insulin secretion.

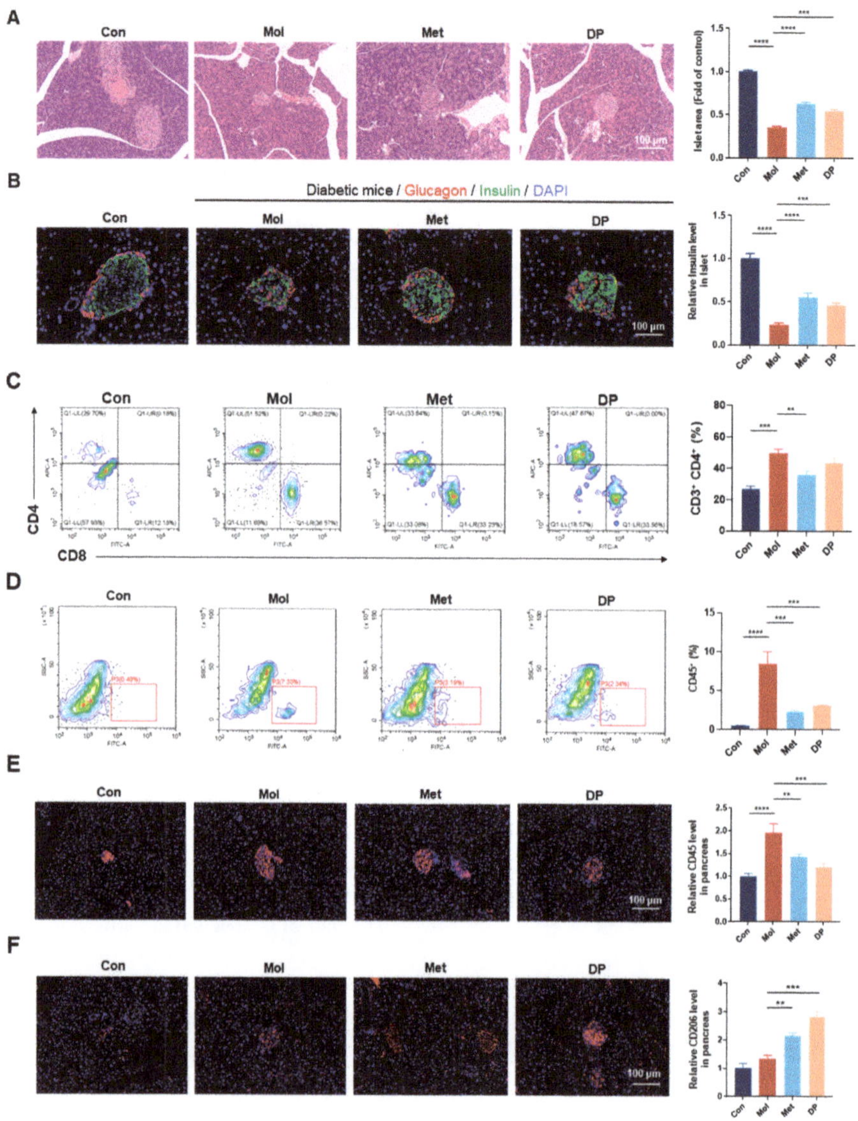

Figure 6. Effect of DP on pancreatic histopathology and in vivo anti-T2DM immune response of DP. (**A**) Representative images of HE-stained pancreas section. (**B**) Pancreatic insulin (Green) and glucagon (Red) CLSM. (**C**) Flow cytometric analysis of CD4+ and CD8+ cells gating on CD3+ cells in the pancreas. (**D**) Flow cytometric analysis of CD45 cells in the pancreas. (**E**) Representative CSLM images of CD45 cells (Red) in the pancreas. (**F**) Representative CSLM images of CD206 cells (Red) in the pancreas. The percentages of populations of cells in each group are presented as histograms. Scale bar = 100 μm (n = 3, data are presented as mean ± SD). Statistical significances between every two groups were calculated via one-way ANOVA. ** $p < 0.01$, *** $p < 0.001$, **** $p < 0.0001$.

3.6. Effects of DP on Islet Inflammation Regulation in Mice with T2DM

To investigate the impact of DP intervention on immune inflammation in the islets of T2DM mice, we analyzed the expression of activated T-cell markers (CD4+ and CD8+) in pancreatic tissue of HFD/STZ-induced T2DM mice using flow cytometry (Figure 6C).

Compared to the Con group, the Mol group exhibited a higher proportion of CD4+ cells ($p < 0.01$), accounting for 49.7%, confirming that T2DM promotes inflammatory progression. Following metformin and DP interventions, the proportion of CD4+ cells in the Met group significantly decreased to 36.13% ($p < 0.01$). The DP group also showed a reduction in CD4+ cells to 43.32%, although this was not statistically significant. These results suggested that DP may enhance immune regulation to some extent.

Figure 6D shows the flow cytometry results for CD45 in mouse pancreatic tissue. Compared to the Con group, the proportion of CD45+ cells in T2DM mice was significantly increased ($p < 0.01$). After DP intervention, the CD45+ ratio decreased from 8.44% to 3.12% ($p < 0.01$), while metformin intervention reduced the CD45+ ratio to 2.36% ($p < 0.01$). Similarly, immunofluorescence analysis of mouse pancreatic tissue showed a consistent trend (Figure 6E). These results indicated that DP could significantly suppress pancreatic tissue inflammation and reduce CD45+ cell infiltration.

To explore the effects of DP on macrophages, we performed immunofluorescence staining using CD206 to label M2 macrophages (red) and DAPI to label nuclei (blue) (Figure 6F). The results revealed that compared to the Con group, the expression of CD206+ in the pancreas of T2DM mice was significantly increased ($p < 0.01$). Most importantly, after DP intervention, the expression of CD206+ in the pancreas increased by 2.09-fold compared to the Mol group and was significantly higher than in the Met group. These findings suggested that the infiltration of M2-type macrophages might be associated with the alleviation of local pancreatic inflammation, but its causality needs to be further verified.

3.7. *Effects of DP on Gut Microbiota in T2DM Mice*

We conducted 16S rDNA sequencing on fecal samples from each group of mice to determine the impact of DP on the composition of the gut microbiota. To identify taxa unique to different groups, a Venn diagram was used to cluster and partition operational taxonomic units (OTUs) at a 97% sequence similarity threshold. A total of 5064 OTUs were identified across all groups (Figure 7A). Among these, 1774, 1819, 1869, and 1750 OTUs were observed in the control group (Con), model group (Mol), metformin group (Met), and DP group, respectively. The DP group had 1750 OTUs, with 882 being unique.

For β-diversity analysis, intergroup differences analysis and principal coordinate analysis (PCoA) based on Unweighted Unifrac distances were performed (Figure 7B). The results showed that the microbial community structure was completely separated into two distinct clusters, with samples within each cluster being closely distributed. Intergroup differences were greater than intragroup differences ($p < 0.01$), indicating that HFD/STZ induction and DP intervention significantly altered the composition of the gut microbiota.

Additionally, linear discriminant analysis effect size (LEfSe) was used to highlight core bacterial phenotypes from phylum to genus that contributed to changes in microbiota composition (Figure 7C,D). At the phylum level, the Con group was enriched with *Bacteroidia* and *Parabacteroides*. At the family level, *Muribaculaceae, Prevotellaceae, Tannerellaceae, Sutterellaceae, Marinfilaceae, Clostridiaceae*, and *Micrococcaceae* were enriched. At the genus level, *Alloprevotella, Parabacteria*, and *Parasutterella* were enriched. Thus, these results suggested the potential role of DP in modulating microbiota abundance.

In the Mol group, *Actinobacteriota, Verrucomicrobiae*, and *Cyanobacteria* were enriched at the phylum level, while *Faecalibaculum, Dubosiella, Aerococcus, Turicibacter*, and *Romboutsia* were enriched at the genus level. These findings indicated that HFD/STZ induction reduced the enrichment of gut microbiota observed in the Con group. In the DP group, *Lactobacillaceae* was enriched at the family level, and *Lactobacillus, Desulfovibrio*, and *Parvibacter* were enriched at the genus level.

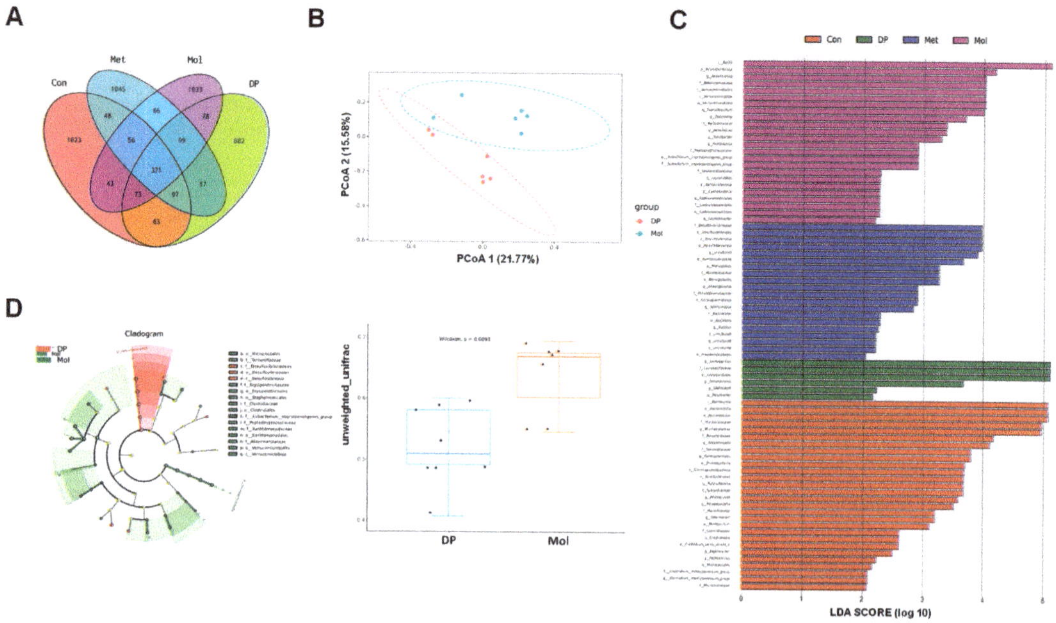

Figure 7. Effects of DP on the diversity of gut microbiota in mice with T2DM. (**A**) OTUs Venn graph of mice feces in each group. (**B**) Inter-group difference analysis of β-diversity based on Unweighted Unifrac distance and PCoA analysis diagram. (**C**) LDA scores of taxa enriched at different taxonomy levels. (**D**) Taxonomic cladogram generated by LEfSe analysis (n = 6).

4. Discussion

Type 2 diabetes mellitus (T2DM) is a metabolic disorder characterized by chronic hyperglycemia due to insulin resistance in peripheral tissues and progressive dysfunction of β-cells in the pancreas. Insulin resistance (IR) is considered a precursor and primary pathogenic factor of T2DM [27]. The liver is a key target for insulin activity, and HepG2 cells are commonly utilized in vitro to study hepatic insulin resistance [28]. The transport of glucose across the plasma membrane is the rate-limiting step for glucose entry into cells. To enhance cellular glucose uptake, insulin signaling promotes the translocation of specific glucose transporters to the plasma membrane [29]. In this study, we induced insulin resistance in HepG2 cells and found that DP intervention significantly improved glucose uptake in the IR-HepG2 model. The IRS1/PI3K/AKT pathway is a major signaling pathway associated with insulin resistance, mediating insulin-stimulated glucose uptake and utilization. Insulin binds to its receptor, triggering the phosphorylation of insulin receptor substrates, which subsequently activates PI3K and its downstream effector AKT [30,31]. Serine phosphorylation of IRS1 can inhibit its tyrosine phosphorylation, blocking downstream pathways and impairing insulin signaling. GLUT4, a major insulin-responsive glucose transporter, can translocate from perinuclear compartments (including GLUT4 storage vesicles) to the plasma membrane [32,33]. To further explore the mechanism of DP, we assessed the levels of IRS1 and GLUT4 in the IR-HepG2 model before and after DP intervention. Our findings indicated that DP promoted glucose uptake by upregulating IRS1 and GLUT4, suggesting that DP may improve glucose metabolism by activating the IRS1/GLUT4 signaling pathway.

Prolonged consumption of a high-fat diet results in excessive fat storage and elevated free fatty acids in diabetic mice, leading to insulin resistance and hindered glucose metabolism [34]. A T2DM mouse model was created through HFD/STZ induction, with metformin serving as a reference treatment. DP intervention significantly reversed typical features of T2DM, including weight loss, elevated FBG, and abnormal oral glucose tolerance. Notably, the Met group exhibited sustained lower body weight, possibly attributed to the drug's adverse effects that impeded weight restoration. Our previous studies have shown that DP can increase skeletal muscle strength and improve muscle mass and strength, which may significantly benefit T2DM patients experiencing weight loss [35]. Additionally, accelerated lipid breakdown during diabetes leads to abnormal levels of total cholesterol (TC) and triglycerides (TG) [36]. In this study, DP intake reduced serum levels of TC, TG, and LDL-C, while increasing levels of TP and ALB in T2DM mice. These findings indicate that DP can regulate lipid and protein metabolism while stabilizing glucose levels, thereby mitigating diabetic complications to some extent.

T2DM is frequently associated with hepatic steatosis, although the exact causal relationship between the two conditions remains uncertain. Hepatic lipid accumulation could serve as a potential indicator of T2DM [37]. Excessive intake of high-sugar and high-fat diets releases large amounts of fatty acids into the liver. When these exceed the liver's clearance capacity, lipid deposition occurs, leading to hepatic steatosis [38]. Hepatic glycogen levels play a crucial role in monitoring glucose metabolism. The liver regulates glucose output and glycogen synthesis to maintain glucose homeostasis. Analysis of liver tissue samples from treated mice indicated a significant improvement in liver damage and an increase in hepatic glycogen content with DP treatment. This effect may be attributed to DP providing protein nutrition to T2DM mice, enabling damaged hepatocytes to engage their repair mechanisms and regain normal structural integrity. Additionally, the enzymes and coenzymes required for various reactions in hepatocytes were sufficiently supplied, allowing rapid restoration of liver function. Insulin regulates blood glucose levels by binding to its receptor and stimulating downstream proteins. T2DM is associated with disturbances in various glucose metabolism-related signaling pathways in the liver, notably the IRS1/PI3K/AKT pathway [29]. In the absence of abnormalities, insulin binding triggers tyrosine phosphorylation of IRS1, suppressing serine phosphorylation and initiating subsequent signaling cascades. Conversely, in insulin resistance, serine phosphorylation of IRS1 intensifies, impeding downstream signaling activation. Following intervention with DP, there was a partial recovery in the levels of IRS1 and PI3K proteins, indicating that DP might ameliorate hepatic insulin resistance by stimulating the IRS1/PI3K signaling pathway.

Islet cell damage is a hallmark of T2DM. Histological analysis of pancreatic tissues showed that DP effectively alleviated islet damage in T2DM mice. In addition to islet cell damage, T2DM disrupts insulin secretion regulation, leading to insufficient insulin production [39]. Our results showed that both metformin and DP interventions significantly promoted insulin secretion, consistent with findings by OGIWARA et al. [40]. Pancreatic inflammation is a critical component of T2DM pathophysiology [41]. In patients with T2DM, the number of macrophages in islets increased significantly, along with elevated levels of cytokines and chemokines. Additionally, increased expression of CD4+ cells and macrophages has been observed in the exocrine compartment [42]. These studies suggested that pancreatic inflammation accompanies T2DM. CD4+ and CD8+ T-cell subsets play a central role in the immune system, maintaining immune balance through mutual regulation. Nevertheless, persistent inflammation during the advancement of T2DM disrupts this equilibrium, resulting in metabolic irregularities and the accumulation of metabolic byproducts, thereby intensifying immune dysfunction and disease progression [43]. CD3+ T-cell activation markers (CD4+ and CD8+) in pancreatic tissues of HFD/STZ-induced T2DM mice analyzed by flow cytometry revealed that DP reduced the proportion of CD4+ cells compared to T2DM mice, although not significantly, suggesting potential immune-modulating effects. During inflammation, the number of immune-related cells increases.

Studies have shown that CD45+ leukocyte infiltration in islets was higher in T2DM patients compared to healthy individuals [44]. Reducing leukocyte infiltration may help alleviate inflammation. Flow cytometry and immunofluorescence analysis showed a significant reduction in CD45+ cell infiltration after DP intervention. Previous studies have highlighted the role of protein nutrition in reducing inflammation, consistent with our findings [45]. These results indicated that DP can significantly suppress pancreatic inflammation and reduced CD45+ infiltration. Macrophages, as innate immune cells, can polarize into M1 or M2 phenotypes under physiological and pathological conditions. T2DM-induced hyperglycemia promoted M1 polarization and inhibited M2 polarization, thus aggravating inflammation and disrupting immune balance [46]. Therefore, redirecting and reshaping macrophage polarization is considered a therapeutic strategy for T2DM. Our findings indicated that DP intervention significantly elevated CD206+ expression in pancreatic tissue. Consequently, we speculated that the infiltration of M2-type macrophages might be associated with the alleviation of local pancreatic inflammation, but its causality needs to be further verified.

As understanding of the gut microbiota deepens, increasing evidence suggests its role in metabolic disorders such as obesity and T2DM. Dysbiosis of the gut microbiota increases intestinal permeability, allowing bacterial endotoxins to enter the bloodstream, triggering inflammation, insulin resistance, and hyperglycemia [47]. To elucidate the effects of DP supplementation on gut microbiota, we examined changes in microbial composition and bacterial relative abundance in treated mice. In this study, DP supplementation altered the gut microbiota composition in T2DM mice. Different bacteria exert varying effects on host homeostasis. Notably, DP increased the abundance of *Lactobacillus*, *Parvibacter*, and *Lactobacillaceae*. The enrichment of *Lactobacillus* may be associated with plant-based protein intake, which is closely linked to the development of chronic metabolic diseases, particularly diabetes [48]. Studies have shown that Lactobacillus treatment can alleviate FBG levels, postprandial glucose, glucose tolerance, and liver damage in HFD/STZ-induced diabetic mice. The anti-diabetic potential of Lactobacillus may involve regulating glucose and lipid metabolism, energy metabolism, and reducing systemic inflammation [49]. *Parvibacter*, a naturally occurring commensal bacterium, constitutes 3–5% of the healthy human gut microbiome. Numerous microbiome analyses have shown reduced *Parvibacter* abundance in various disease states, including irritable bowel disease, Crohn's disease, asthma, depression, and metabolic disorders [50]. Specifically, reduced *Parvibacter* levels are associated with elevated blood glucose and T2DM. To sum up, our findings indicated that DP could prevent T2DM via repairing gut microbiota.

5. Conclusions

In this study, we investigated the effects of DP on the HepG2 insulin resistance cell model at the in vitro level. The findings demonstrated that DP increased glucose uptake by potentially modulating the IRS1/GLUT4 signaling pathway, thus alleviating insulin resistance in HepG2 cells. In the T2DM mice model, nutritional intervention with DP significantly enhanced glucose, lipid, and protein metabolism, reduced inflammation in HFD/STZ-induced T2DM mice, and improved liver and pancreatic function. These effects were potentially mediated through the activation of the Insulin/IRS1/PI3K signaling pathway and modulation of gut microbiota composition and abundance. In conclusion, these results suggest that soy–whey dual-protein supplementation may ameliorate hepatic insulin resistance and T2DM, offering a novel and promising nutritional intervention for T2DM patients to mitigate inflammatory responses and protect organ function. Further studies will be conducted in the future, including dose-response studies to optimize therapeutic regimes, metabolic profiling to clarify microbiota-mediated mechanisms, and explorations of alternative pathways. In addition, targeted clinical trials can also provide more powerful evidence for the application of dual-protein in the prevention and treatment of diabetes.

Supplementary Materials: The following supporting information can be downloaded at: https://www.mdpi.com/article/10.3390/foods14122115/s1, Table S1: Amino Acid Profile of Dual-Protein.

Author Contributions: N.L.: Conceptualization, Investigation, Writing—original draft. H.L.: Investigation. D.F.: Data analysis, Investigation. M.L.: Writing—review and editing, Supervision. D.H.: Writing—review and editing, Investigation. T.L.: Supervision. J.W.: Conceptualization, Resources, Supervision. All authors have read and agreed to the published version of the manuscript.

Funding: This research was funded by the Innovation Engineering Project of the Chinese Academy of Agricultural Sciences (CAAS-ASTIP-2025-IFND).

Data Availability Statement: Data is provided within the manuscript information files.

Conflicts of Interest: The authors declare that the research was conducted in the absence of any commercial or financial relationships that could be construed as a potential conflict of interest.

Abbreviation

DP, dual-protein; T2DM, type 2 diabetes mellitus; HFD, high-fat diet; STZ, streptozotocin; IR, Insulin resistance; IRS1, Insulin receptor substrate 1; GLUT4, Glucose transporter 4; CLSM, Confocal laser scanning microscopy; FBG, Fasting blood glucose; INS, Insulin; LDL, Low density lipoprotein; OGTT, Oral glucose tolerance test; PBS, Phosphate buffered solution; PI3K, Phosphatidylin-ositol-3-kinase; TC, Total cholesterol; TG, Triglyceride; TP, Total protein; ALB, Albumin.

References

1. Cole, J.B.; Florez, J.C. Genetics of diabetes mellitus and diabetes complications. *Nat. Rev. Nephrol.* **2020**, *16*, 377–390. [CrossRef] [PubMed]
2. Bailey, C.J. Metformin: Historical overview. *Diabetologia* **2017**, *60*, 1566–1576. [CrossRef]
3. Hamada, Y.; Nagasaki, H.; Fuchigami, M.; Furuta, S.; Seino, Y.; Nakamura, J.; Oiso, Y. The alpha-glucosidase inhibitor miglitol affects bile acid metabolism and ameliorates obesity and insulin resistance in diabetic mice. *Metabolism* **2013**, *62*, 734–742. [CrossRef]
4. Dennis, J.M.; Henley, W.E.; Weedon, M.N.; Lonergan, M.; Rodgers, L.R.; Jones, A.G.; Hamilton, W.T.; Sattar, N.; Janmohamed, S.; Holman, R.R.; et al. Sex and BMI Alter the Benefits and Risks of Sulfonylureas and Thiazolidinediones in Type 2 Diabetes: A Framework for Evaluating Stratification Using Routine Clinical and Individual Trial Data. *Diabetes Care* **2018**, *41*, 1844–1853. [CrossRef] [PubMed]
5. Malaeb, S.; Bakker, C.; Chow, L.S.; Bantle, A.E. High-Protein Diets for Treatment of Type 2 Diabetes Mellitus: A Systematic Review. *Adv. Nutr.* **2019**, *10*, 621–633. [CrossRef]
6. Layman, D.K.; Clifton, P.; Gannon, M.C.; Krauss, R.M.; Nuttall, F.Q. Protein in optimal health: Heart disease and type 2 diabetes. *Am. J. Clin. Nutr.* **2008**, *87*, 1571S–1575S. [CrossRef]
7. Chalvon-Demersay, T.; Azzout-Marniche, D.; Arfsten, J.; Egli, L.; Gaudichon, C.; Karagounis, L.G.; Tomé, D. A Systematic Review of the Effects of Plant Compared with Animal Protein Sources on Features of Metabolic Syndrome. *J. Nutr.* **2017**, *147*, 281–292. [CrossRef] [PubMed]
8. Stevenson, E.J.; Allerton, D.M. The role of whey protein in postprandial glycaemic control. *Proc. Nutr. Soc.* **2018**, *77*, 42–51. [CrossRef]
9. Das, D.; Kabir, M.E.; Sarkar, S.; Wann, S.B.; Kalita, J.; Manna, P. Antidiabetic potential of soy protein/peptide: A therapeutic insight. *Int. J. Biol. Macromol.* **2022**, *194*, 276–288. [CrossRef]
10. Woo, H.W.; Kim, M.K.; Lee, Y.H.; Shin, D.H.; Shin, M.H.; Choi, B.Y. Sex-specific associations of habitual intake of soy protein and isoflavones with risk of type 2 diabetes. *Clin. Nutr.* **2021**, *40*, 127–136. [CrossRef]
11. Baer, D.J.; Stote, K.S.; Paul, D.R.; Harris, G.K.; Rumpler, W.V.; Clevidence, B.A. Whey protein but not soy protein supplementation alters body weight and composition in free-living overweight and obese adults. *J. Nutr.* **2011**, *141*, 1489–1494. [CrossRef]
12. Ji, A.; Chen, W.; Zhang, T.; Shi, R.; Wang, X.; Wang, Y.; Xu, H.; Li, D. Whey protein and soy protein prevent obesity by upregulating uncoupling protein 1 to activate brown adipose tissue and promote white adipose tissue browning in high-fat diet-fed mice. *Food Funct.* **2022**, *13*, 12836–12851. [CrossRef] [PubMed]

13. Gobert, C.P.; Pipe, E.A.; Capes, S.E.; Darlington, G.A.; Lampe, J.W.; Duncan, A.M. Soya protein does not affect glycaemic control in adults with type 2 diabetes. *Br. J. Nutr.* **2010**, *103*, 412–421. [CrossRef] [PubMed]
14. Sambashivaiah, S.; Cope, M.; Mukherjea, R.; Selvam, S.; George, N.; Kuriyan, R.; Kurpad, A.V.; Sweazea, K.L. The Effect of Soy and Whey Protein Supplementation on Glucose Homeostasis in Healthy Normal Weight Asian Indians. *J. Nutr. Metab.* **2023**, *2023*, 2622057. [CrossRef] [PubMed]
15. Piri Damaghi, M.; Mirzababaei, A.; Moradi, S.; Daneshzad, E.; Tavakoli, A.; Clark, C.C.T.; Mirzaei, K. Comparison of the effect of soya protein and whey protein on body composition: A meta-analysis of randomised clinical trials. *Br. J. Nutr.* **2022**, *127*, 885–895. [CrossRef]
16. Guyomarc'h, F.; Arvisenet, G.; Bouhallab, S.; Canon, F.; Deutsch, S.M.; Drigon, V.; Dupont, D.; Famelart, M.-H.; Garric, G.; Guédon, E.; et al. Mixing milk, egg and plant resources to obtain safe and tasty foods with environmental and health benefits. *Trends Food Sci. Technol.* **2021**, *108*, 119–132. [CrossRef]
17. Wu, C.; Wang, T.; Ren, C.; Ma, W.; Wu, D.; Xu, X.; Wang, L.; Du, M. Advancement of food-derived mixed protein systems: Interactions, aggregations, and functional properties. *Compr. Rev. Food Sci. Food Saf.* **2021**, *20*, 627–651. [CrossRef]
18. Yang, J.Q.; Liu, G.Y.; Zeng, H.B.; Chen, L.Y. Effects of high pressure homogenization on faba bean protein aggregation in relation to solubility and interfacial properties. *Food Hydrocoll.* **2018**, *83*, 275–286. [CrossRef]
19. Wu, M.; He, X.; Feng, D.; Li, H.; Han, D.; Li, Q.; Zhao, B.; Li, N.; Liu, T.; Wang, J. The Effect of High-Pressure Homogenization on the Structure of Dual-Protein and Its Emulsion Functional Properties. *Foods* **2023**, *12*, 3358. [CrossRef]
20. Nell, S.; Suerbaum, S.; Josenhans, C. The impact of the microbiota on the pathogenesis of IBD: Lessons from mouse infection models. *Nat. Rev. Microbiol.* **2010**, *8*, 564–577. [CrossRef]
21. Yan, F.; Li, N.; Yue, Y.; Wang, C.; Zhao, L.; Evivie, S.E.; Li, B.; Huo, G. Screening for Potential Novel Probiotics with Dipeptidyl Peptidase IV-Inhibiting Activity for Type 2 Diabetes Attenuation in vitro and in vivo. *Front. Microbiol.* **2020**, *10*, 2855. [CrossRef] [PubMed]
22. Wu, S.; Bhat, Z.F.; Gounder, R.S.; Mohamed Ahmed, I.A.; Al-Juhaimi, F.Y.; Ding, Y.; Bekhit, A.E. Effect of Dietary Protein and Processing on Gut Microbiota—A Systematic Review. *Nutrients* **2022**, *14*, 453. [CrossRef]
23. Ren, G.; Zhang, J.; Li, M.; Yi, S.; Xie, J.; Zhang, H.; Wang, J. Protein blend ingestion before allogeneic stem cell transplantation improves protein-energy malnutrition in patients with leukemia. *Nutr. Res.* **2017**, *46*, 68–77. [CrossRef] [PubMed]
24. Zhang, J.; Zhang, Q.; Liu, H.; Liu, X.; Yu, Y.; Han, D.; He, X.; Zeng, P.; Wang, J. Soy-whey dual-protein alleviates osteoporosis of ovariectomized rats via regulating bone fat metabolism through gut-liver-bone axis. *Nutrition* **2022**, *103–104*, 111723. [CrossRef]
25. Zheng, X.; Ke, Y.; Feng, A.; Yuan, P.; Zhou, J.; Yu, Y.; Wang, X.; Feng, W. The Mechanism by Which Amentoflavone Improves Insulin Resistance in HepG2 Cells. *Molecules* **2016**, *21*, 624. [CrossRef] [PubMed]
26. Niu, F.; Liu, W.; Ren, Y.; Tian, Y.; Shi, W.; Li, M.; Li, Y.; Xiong, Y.; Qian, L. β-cell neogenesis: A rising star to rescue diabetes mellitus. *J. Adv. Res.* **2024**, *62*, 71–89. [CrossRef]
27. Lu, B.; Yang, Y.; Yang, Z.; Feng, X.; Wang, X.; Zhang, Z.; Hu, R. Insulin resistance in Chinese patients with type 2 diabetes is associated with C-reactive protein independent of abdominal obesity. *Cardiovasc. Diabetol.* **2010**, *9*, 92. [CrossRef]
28. Petersen, M.C.; Shulman, G.I. Mechanisms of Insulin Action and Insulin Resistance. *Physiol. Rev.* **2018**, *98*, 2133–2223. [CrossRef]
29. Ter Horst, K.W.; Gilijamse, P.W.; Ackermans, M.T.; Soeters, M.R.; Nieuwdorp, M.; Romijn, J.A.; Serlie, M.J. Impaired insulin action in the liver, but not in adipose tissue or muscle, is a distinct metabolic feature of impaired fasting glucose in obese humans. *Metabolism* **2016**, *65*, 757–763. [CrossRef]
30. Huang, X.; Liu, G.; Guo, J.; Su, Z. The PI3K/AKT pathway in obesity and type 2 diabetes. *Int. J. Biol. Sci.* **2018**, *14*, 1483–1496. [CrossRef]
31. James, D.E.; Stöckli, J.; Birnbaum, M.J. The aetiology and molecular landscape of insulin resistance. *Nat. Rev. Mol. Cell Biol.* **2021**, *22*, 751–771. [CrossRef] [PubMed]
32. Alam, F.; Islam, M.A.; Khalil, M.I.; Gan, S.H. Metabolic Control of Type 2 Diabetes by Targeting the GLUT4 Glucose Transporter: Intervention Approaches. *Curr. Pharm. Des.* **2016**, *22*, 3034–3049. [CrossRef] [PubMed]
33. Minami, S.; Yokota, N.; Kawahara, H. BAG6 contributes to glucose uptake by supporting the cell surface translocation of the glucose transporter GLUT4. *Biol. Open* **2020**, *9*, bio047324. [CrossRef]
34. Kowalski, G.M.; Hamley, S.; Selathurai, A.; Kloehn, J.; De Souza, D.P.; O'callaghan, S.; Nijagal, B.; Tull, D.L.; McConville, M.J.; Bruce, C.R. Reversing diet-induced metabolic dysregulation by diet switching leads to altered hepatic de novo lipogenesis and glycerolipid synthesis. *Sci. Rep.* **2016**, *6*, 27541. [CrossRef]
35. Ren, G.; Yi, S.; Zhang, H.; Wang, J. Ingestion of soy-whey blended protein augments sports performance and ameliorates exercise-induced fatigue in a rat exercise model. *Food Funct.* **2017**, *8*, 670–679. [CrossRef] [PubMed]
36. Bahiru, E.; Hsiao, R.; Phillipson, D.; Watson, K.E. Mechanisms and Treatment of Dyslipidemia in Diabetes. *Curr. Cardiol. Rep.* **2021**, *23*, 26. [CrossRef]

37. Sehgal, R.; Jähnert, M.; Lazaratos, M.; Speckmann, T.; Schumacher, F.; Kleuser, B.; Ouni, M.; Jonas, W.; Schürmann, A. Altered liver lipidome markedly overlaps with human plasma lipids at diabetes risk and reveals adipose-liver interaction. *J. Lipid Res.* **2025**, *66*, 100767. [CrossRef]
38. Westcott, F.; Dearlove, D.J.; Hodson, L. Hepatic fatty acid and glucose handling in metabolic disease: Potential impact on cardiovascular disease risk. *Atherosclerosis* **2024**, *394*, 117237. [CrossRef]
39. Aguayo-Mazzucato, C.; Bonner-Weir, S. Pancreatic β Cell Regeneration as a Possible Therapy for Diabetes. *Cell Metab.* **2018**, *27*, 57–67. [CrossRef]
40. Ogiwara, M.; Ota, W.; Mizushige, T.; Kanamoto, R.; Ohinata, K. Enzymatic digest of whey protein and wheylin-1, a dipeptide released in the digest, increase insulin sensitivity in an Akt phosphorylation-dependent manner. *Food Funct.* **2018**, *9*, 4635–4641. [CrossRef]
41. Donath, M.Y.; Böni-Schnetzler, M.; Ellingsgaard, H.; Ehses, J.A. Islet inflammation impairs the pancreatic beta-cell in type 2 diabetes. *Physiology* **2009**, *24*, 325–331. [CrossRef] [PubMed]
42. Sarikonda, G.; Pettus, J.; Phatak, S.; Sachithanantham, S.; Miller, J.F.; Wesley, J.D.; Cadag, E.; Chae, J.; Ganesan, L.; Mallios, R.; et al. CD8 T-cell reactivity to islet antigens is unique to type 1 while CD4 T-cell reactivity exists in both type 1 and type 2 diabetes. *J. Autoimmun.* **2014**, *50*, 77–82. [CrossRef]
43. de Candia, P.; Prattichizzo, F.; Garavelli, S.; De Rosa, V.; Galgani, M.; Di Rella, F.; Spagnuolo, M.I.; Colamatteo, A.; Fusco, C.; Micillo, T.; et al. Type 2 Diabetes: How Much of an Autoimmune Disease? *Front. Endocrinol.* **2019**, *10*, 451. [CrossRef]
44. Böni-Schnetzler, M.; Meier, D.T. Islet inflammation in type 2 diabetes. *Semin. Immunopathol.* **2019**, *41*, 501–513. [CrossRef] [PubMed]
45. Prokopidis, K.; Mazidi, M.; Sankaranarayanan, R.; Tajik, B.; McArdle, A.; Isanejad, M. Effects of whey and soy protein supplementation on inflammatory cytokines in older adults: A systematic review and meta-analysis. *Br. J. Nutr.* **2023**, *129*, 759–770. [CrossRef] [PubMed]
46. Hu, N.; Zhang, X.; Zhang, X.; Guan, Y.; He, R.; Xue, E.; Deng, W.; Yu, J.; Wang, W.; Shi, Q. Inhibition of Notch activity suppresses hyperglycemia-augmented polarization of macrophages to the M1 phenotype and alleviates acute pancreatitis. *Clin. Sci.* **2022**, *136*, 455–471. [CrossRef]
47. Gueddouri, D.; Caüzac, M.; Fauveau, V.; Benhamed, F.; Charifi, W.; Beaudoin, L.; Rouland, M.; Sicherre, F.; Lehuen, A.; Postic, C.; et al. Insulin resistance per se drives early and reversible dysbiosis-mediated gut barrier impairment and bactericidal dysfunction. *Mol. Metab.* **2022**, *57*, 101438. [CrossRef]
48. Wei, Z.; Zhou, N.; Zou, L.; Shi, Z.; Dun, B.; Ren, G.; Yao, Y. Soy Protein Alleviates Malnutrition in Weaning Rats by Regulating Gut Microbiota Composition and Serum Metabolites. *Front. Nutr.* **2021**, *8*, 774203. [CrossRef]
49. Wang, G.; Si, Q.; Yang, S.; Jiao, T.; Zhu, H.; Tian, P.; Wang, L.; Li, X.; Gong, L.; Zhao, J.; et al. Lactic acid bacteria reduce diabetes symptoms in mice by alleviating gut microbiota dysbiosis and inflammation in different manners. *Food Funct.* **2020**, *11*, 5898–5914. [CrossRef]
50. Pang, B.; Jin, H.; Liao, N.; Li, J.; Jiang, C.; Shi, J. Vitamin A supplementation ameliorates ulcerative colitis in gut microbiota-dependent manner. *Food Res. Int.* **2021**, *148*, 110568. [CrossRef]

foods

MDPI

Communication

Enhanced Protein Extraction from *Auxenochlorella protothecoides* Through Synergistic Mechanical Cell Disruption and Alkaline Solubilization

Jun Wei Ng, Sze Ying Lee, Tong Mei Teh, Melanie Weingarten and Md. Mahabubur Rahman Talukder *

Singapore Institute of Food and Biotechnology Innovation (SIFBI), Agency for Science, Technology, and Research (A*STAR), 31 Biopolis Way, Nanos, Singapore 138669, Singapore; ng_jun_wei@a-star.edu.sg (J.W.N.); lee_sze_ying@a-star.edu.sg (S.Y.L.); teh_tong_mei@a-star.edu.sg (T.M.T.); melanie_weingarten@a-star.edu.sg (M.W.)

* Correspondence: rahman_talukder@a-star.edu.sg

Abstract

Microalgae proteins are increasingly recognized in the food and nutraceutical industries for their functional versatility and high nutritional value. Mild alkaline treatment is commonly used for cell wall degradation and intracellular protein solubilization, consequently enhancing the protein extraction yield. The findings of this study reveal that alkaline treatment alone, even at higher NaOH concentration (up to 0.3 M) and treatment time (up to 90 min), was ineffective (max. 2.4% yield) for the extraction of protein from *Auxenochlorella protothecoides* biomass. This challenge was significantly reduced through synergistic application of mechanical cell disruption using high-pressure homogenization (HPH) and alkaline solubilization. Single-pass HPH (35 k psi) alone without alkaline treatment led to 52.3% protein solubilization from wet biomass directly harvested from culture broth, while it was only 18.5% for spray-dried biomass. The combined effect of HPH and alkaline (0.1 M NaOH) treatment significantly increased protein extraction yield to 68.0% for a spray-dried biomass loading of 50 g L^{-1}. Through replacing spray-dried biomass with wet biomass, the requirement of NaOH was reduced by 5-fold to 0.02 M to achieve a similar yield of 68.1%. The process integration of HPH with the mild alkaline solubilization and utilization of wet biomass from culture broth showed high potential for industrialization of microalgae protein extraction. This method achieves high extraction yield while reducing alkaline waste and eliminating the need for energy-consuming drying of biomass, thereby minimizing the environmental impact.

Keywords: microalgae; wet biomass processing; mechanical cell disruption; high-pressure homogenization; alkaline solubilization; extraction

1. Introduction

Single-cell protein (SCP) or protein-rich microbial biomass is increasingly important due to its sustainability, nutritional value, and potential applications in food, feed, and biotechnology. While yeast/fungus and bacteria remain valuable SCP sources due to their fast growth, microalgae offer superior sustainability, high protein content (50–70%) [1], broader nutritional diversity including bioactive, and a lower environmental impact. Additionally, they utilize sunlight and CO_2, making them a highly attractive alternative for future protein production. However, the rigid cell wall of microalgae like *Chlorella vulgaris* and limited solubility of microalgal protein in water pose a significant challenge, leading to low intracellular component availability and hindering efficient protein extraction [2]. Hence, an effective method for cell disruption and intracellular protein solubilization is a prerequisite to preparing microalgae protein isolate/concentrate. Furthermore, such a method should not only maximize protein recovery yield but must be scalable, cost-effective, and avoid degradation of the protein. The protein degradation or hydrolysis affects the protein attributes (e.g., increases bitter taste and alters protein functionalities). Elevated NaOH concentrations (pH > 12) can lead to protein hydrolysis, reducing functional properties [3].

A chemical approach using very mild alkaline water (pH up to 9.5) is commonly used in the industrial extraction of plant proteins, such as soy protein isolates or concentrates. This process permeabilizes plant cell walls, solubilizes proteins, and enables subsequent recovery via isoelectric precipitation and drying. However, our recent findings suggest that the mild alkaline treatment (pH 9.0) is ineffective in permeabilizing the rigid cell wall of C. vulgaris, resulting in limited protein solubilization (yield 6%) at a biomass load of 50 g L^{-1} [4]. Safi et al. [5] also reported a modest protein recovery of 26% from *C. vulgaris* using alkaline treatment at an elevated pH of 12 for 24 h for a biomass load of 20 g L^{-1}. The authors found that the sodium hydroxide solution was insufficient to completely hydrolyze the cell wall of *C. vulgaris* but partially weakened it, allowing the release of only small-sized cytoplasmic proteins while hindering the diffusion of larger proteins. We also observed that increasing NaOH concentration enhances protein solubilization; however, a higher NaOH level (>0.1 M) leads to degradation of *Chlorella* protein [4]. Additionally, higher NaOH levels contribute to increased sodium content in the final product, which may be undesirable for food applications.

Numerous studies on physical cell disruption using bead milling, HPH, or ultrasonication have been reported for enhancing microbial cell permeabilization, resulting in higher protein extraction. A study by Postma et al. [6] demonstrated that bead milling resulted in more than 97% cell disintegration, yielding up to 42% of water-soluble proteins from *C. vulgaris*. Safi et al. [5] reported that HPH rapidly disrupted the cells, allowing the recovery of 66% of total proteins from *C. vulgaris*. Jubeau et al. [7] optimized the HPH process operating pressure and number of passages for the selective extraction of proteins and pigments from *Porphyridium cruentum*. Perez et al. [8] enhanced microalgae protein extraction using emerging cell disruption technology, pulsed electric field (PEF). Although PEF could enhance the extraction yield from 11 to 25%, the improved yield is not that high and is lower than HPH. Katsimichas et al. [9] studied the kinetics of HPH-assisted protein extraction and reported protein recovery of 38% after 24 h through aqueous extraction. While previous studies on HPH and bead milling showed promising results for microalgae cell disruption, they did not compare the efficacy of these two methods in the context of both cell disruption and subsequent alkaline protein solubilization. More importantly, these studies did not focus on fine-tuning the concentration of alkaline, a critical factor that not only influences protein solubilization but can also lead to protein degradation and elevated waste generation. Furthermore, the prior work primarily used dried microalgae biomass. The impact of HPH and bead milling on protein solubilization using wet biomass harvested directly from fermentation broth remains unexplored. The utilization of wet biomass eliminates not only the energy-intensive drying step but also simplifies the overall microalgal biomass production process.

This study aims to develop a physico-chemical protein extraction process from wet microalgae biomass to avoid energy-consuming drying and reduce waste generation, an unfulfilled need to prepare microbial protein isolate/concentrate at an industrial scale. Our primary focus was on developing and optimizing a scalable extraction method, and we prioritized yield as initial performance indicator. The reduction of alkaline loading, which affects the protein quality, is another objective of this study. Heterotrophically, lab-grown *A. protothecoides* biomass obtained from cultured broth in a bioreactor was used to optimize HPH and bead milling to compare their effectiveness. Alkaline concentration was fine-tuned to maximize protein solubilization and extraction yield. The spray-dried *Auxenochlorella* biomass served as a control to evaluate the benefits of utilizing wet biomass in terms of protein extraction yield, reduced process severity, and lower alkaline concentration. *A. protothecoides*, formerly *Chlorella protothecoides*, is a versatile microalga capable of autotrophic, mixotrophic, and heterotrophic growth. With a protein content of approximately 50%, it is a promising sustainable protein source. However, research on its protein extraction methods remains limited compared to other microalgae strains like *C. vulgaris*.

2. Materials and Methods

2.1. Chemicals and Reagents

Sodium hydroxide pellet, yeast extract, glucose, Bradford reagent, and bovine serum albumin (BSA) were supplied by Sigma-Aldrich (St. Louis, MO, USA), while all other chemicals used in the study were of analytical grade. All the materials were used without further purification. The water used throughout the work was ultra-pure water treated by a Milli-Q integral water purification system.

2.2. Microalgae Cultivation

The microalgae selected was *A. protothecoides* SAG 211-7a, obtained from the University of Göttingen. A modified Bold's Basal Medium (BBM) in which 6.7 g L^{-1} of yeast extract replaced $NaNO_3$ was prepared for microalgae cultivation, with the composition (g L^{-1}): yeast extract, 6.7; $CaCl_2.2H_2O$, 0.025; $MgSO_4.7H_2O$, 0.075; K_2HPO_4, 0.075; KH_2PO_4, 0.175; NaCl, 0.025; H_3BO_3, 0.114; ethylenediaminetetraacetic acid, 0.05; KOH, 0.031; $FeSO_4.7H_2O$, 0.00498; $ZnSO_4.7H_2O$, 0.0088; $MnCl_2.4H_2O$, 0.0014; MoO_3, 0.0007; $CuSO_4.5H_2O$, 0.0016; and $Co(NO_3)_2.6H_2O$, 0.0005. For liquid precultures, microalgae were incubated in the modified BBM with 20 g L^{-1} glucose for 30 h in a shake flask under heterotrophic conditions (200 rpm, 25 °C, in darkness). These precultures were used to inoculate the 10 L bioreactor cultures with the modified BBM containing 48 g L^{-1} glucose and 40 g L^{-1} yeast extract. The cultivation was performed under batch mode in a Biostat B-DCU 10 L bioreactor (Sartorius, Göttingen, Germany) with a 10% (*v*/*v*) inoculum from the pre-culture. The cultivation conditions applied were a temperature of 25 °C, pH 6.5, and dissolved oxygen maintained at 50% with a cascade system under dark conditions. The cultivation was completed when the residual glucose was depleted.

2.3. Protein Extraction from A. protothecoides

2.3.1. Preparation of Microalgae Biomass

Biomass was harvested at the stationary growth phase of the microalgae. The broth was centrifuged at 3000× *g* for 10 min using a Sorvall Lynx 4000 centrifuge (Thermo Fisher Scientific, Waltham, MA, USA) to remove the liquid content. The biomass pellet obtained was either used directly or spray-dried for protein extraction. The spray-dried conditions were inlet temperature 190 °C, outlet temperature 100–105 °C, feed pump rate 12 mL min^{-1}, and air blower 30 Hz using YC-015 Nano Spray Dryer (Pilotech, Shanghai, China). The protein content of *A. protothecoides* biomass was determined using the Dumas combustion method in a DUMATHERM N Pro analyzer (Gerhardt Analytical Systems, Königswinter, Germany). A nitrogen-to-protein conversion factor of 6.25 was used [10], and the protein content of both wet and spray-dried biomass was 43% on a dry basis.

2.3.2. Protein Extraction Using Alkaline Extraction Alone

Alkaline extraction was applied to wet and spray-dried biomass to obtain microalgae protein. The biomass (5% dry matter) was mixed with 5 mL of sodium hydroxide (NaOH) while stirring at 400 rpm and 37 °C. Different NaOH concentrations (0 to 0.3 M) and mixing times (5 to 90 min) were tested to evaluate their effects on the solubilization yield of the microalgae protein. After mixing, the solution was centrifuged at 3000× *g* for 10 min, and the protein content of the supernatant was determined using the Bradford assay, as stated in Section 2.4. The solubilization yield (%) was calculated using Equation (1):

$$\text{Protein solubilization yield } (\%) = \frac{\text{Solubilized protein in supernatant}}{\text{Total protein in biomass}} \times 100 \quad (1)$$

2.3.3. Protein Extraction Using High-Pressure Homogenization Followed by Alkaline Treatment

The wet and spray-dried biomass concentrations were adjusted to 5% dry matter (50 g L^{-1}) and subjected to HPH in a CF2 cell disruptor (Constant Systems, Daventry, United Kingdom), followed by alkaline solubilization with NaOH. Initially, the biomass was homogenized at 4 °C up to four passes at 35 k psi, with a constant flow rate of 405 mL min^{-1}. After each pass, samples were collected and centrifuged. The soluble protein content in the supernatant was then measured using the Bradford method (Section 2.4) to evaluate the protein solubilization yield using Equation (1).

Subsequently, the homogenized samples were treated with varying NaOH concentrations (0.00002 to 0.3 M) at 37 °C and stirred at 400 rpm for durations ranging from 5 to 60 min. The mixtures were centrifuged to obtain the supernatant, and their protein solubilization yield was determined using Equation (1).

2.3.4. Protein Extraction Using Bead Milling Followed by Alkaline Treatment

The concentration of both wet and spray-dried biomasses was adjusted to 5% dry matter (50 g L^{-1}) in the water. For each treatment, 20 mL of the biomass suspension (5% dry matter) and 20 mL of glass beads were transferred into a 50 mL Falcon tube. The samples were then subjected to bead milling using a Retsch Mixer Mill MM 400 (Retsch, Haan, Germany) under the following conditions: frequency of 26 Hz for 6 cycles of 10 min each, with a 5-min break between cycles. This milling frequency was selected based on our previous findings, which showed that 26 Hz was the most effective in enhancing protein solubilization from dried *C. vulgaris* biomass [11]. The same condition was applied here to assess its effectiveness in disrupting *A. protothecoides* cells and releasing intracellular proteins under varying alkaline conditions.

Following milling, the glass beads were separated, and the resulting biomass slurry was collected. NaOH was added to the biomass slurry to achieve final concentrations ranging from 0.005 to 0.2 M. To solubilize the protein, the mixture was stirred at 400 rpm and 37 °C for 5 min and centrifuged. The protein content in the supernatant before and after alkaline treatment was analyzed using the Bradford assay (Section 2.4), and solubilization yield was estimated based on Equation (1).

2.4. Determination of Soluble Protein Content

The soluble proteins in the supernatant were quantified using the Bradford method [12], as used by other researchers for microalgae protein [13]. A total of 50 µL of sample was mixed with 1.5 mL of Bradford reagent and measured at the wavelength of 595 nm. Protein was quantified using a calibration curve prepared using BSA as a standard.

2.5. Statistics

Each experiment was performed in triplicate under consistent conditions, and all data are expressed as mean ± standard deviation. Error bars in figures represent the standard deviation of mean values. Statistical analysis was conducted using analysis of variance (ANOVA) and Tukey's multiple comparisons in SPSS software version 28.0 (IBM, Armonk, NY, USA) to determine significant differences between mean values at a 95% confidence level ($p < 0.05$).

3. Results and Discussion

3.1. Alkaline Solubilization of Protein

3.1.1. Effect of NaOH Concentration on Protein Solubilization Yield

Both wet and spray-dried *Auxenochlorella* biomass were treated with different alkaline (NaOH) concentrations at a temperature of 37 °C and mixing at 400 rpm for 1 h to understand the effectiveness of breaking the rigid cell wall of *A. protothecoides* and solubilization of intracellular protein through disruption of the complex protein–polysaccharide–lipid structure and deprotonation of acidic amino acids. The results are shown in Figure 1. It can be seen that the protein solubilization yield slightly increased with the increase in NaOH concentrations for both wet and spray-dried biomass, but the solubilization yield was very low (~2.0%), even at elevated concentrations of 0.3 M. It should be noted that after NaOH concentration of 0.1 M, the protein solubilization yield remained the same with the increase in concentration. This result suggests that alkaline treatment alone is not sufficient for breaking the rigid, highly cross-linked cell walls composed of polysaccharides, phospholipid, and glycoprotein [14].

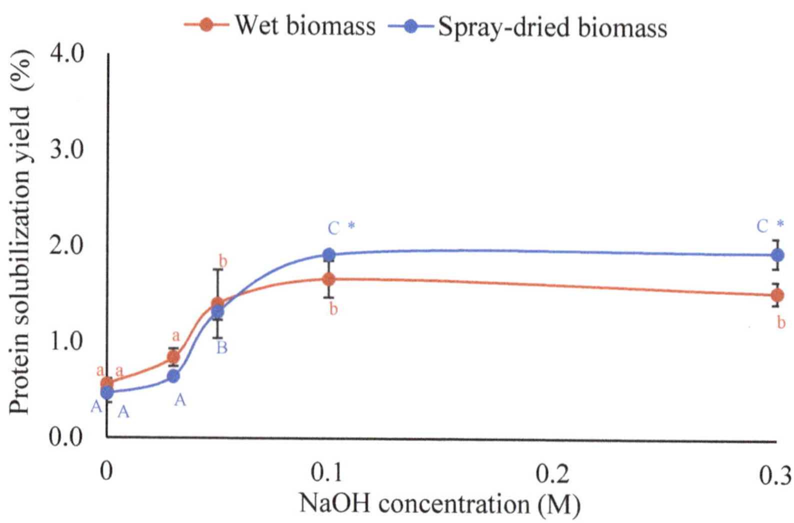

Figure 1. Effect of NaOH concentration on protein solubilization of wet and spray-dried *A. protothecoides* biomass during alkaline treatment. Different lowercase and uppercase letters indicate significant differences between treatments at different NaOH concentrations for wet biomass and spray-dried biomass, respectively ($p < 0.05$). Asterisks (*) indicate significant differences between biomass types (wet vs. spray-dried) at the same NaOH concentration ($p < 0.05$).

3.1.2. Mixing Time for Protein Solubilization

To further investigate the efficacy of alkaline treatment alone on cell breaking and protein solubilization, the mixing time was varied while NaOH concentration was kept constant at 0.1 M NaOH. The NaOH concentration of 0.1 M was chosen based on the result demonstrated in Figure 1. It can be seen from Table 1 that extending the mixing time from 60 to 90 min resulted in only a slight improvement in protein yield, increasing from 1.66% to 2.436% for wet biomass and from 1.915% to 2.274% for spray-dried biomass. Despite this improvement, the overall yields remained low and insignificant, further supporting the conclusion that alkaline treatment alone is insufficient to effectively solubilize microalgal proteins. To enhance protein solubilization, microalgae biomass was physically disrupted using HPH and bead milling. The disrupted biomass was then subjected to alkaline treatment with varying NaOH concentration, as detailed in the following sections.

Table 1. Effect of mixing time on solubilization of wet and spray-dried *A. protothecoides* biomass during alkaline treatment with 0.1 M NaOH. Different lowercase and uppercase letters indicate significant differences between treatments at different mixing times for wet biomass and spray-dried biomass, respectively ($p < 0.05$). Asterisks (*) indicate significant differences between biomass types (wet vs. spray-dried) at the same mixing time ($p < 0.05$).

Mixing Time (min)	Protein Solubilization Yield (%)	
	Wet Biomass	Spray-Dried Biomass
5	0.454 ± 0.005 ᵃ	1.078 ± 0.055 A *
10	0.488 ± 0.035 a	1.419 ± 0.152 B *
20	0.835 ± 0.056 b	1.621 ± 0.087 B *
30	1.043 ± 0.039 c	1.662 ± 0.129 C *
40	1.534 ± 0.079 d	1.857 ± 0.038 CD *
50	1.600 ± 0.086 d	1.872 ± 0.145 D *
60	1.881 ± 0.144 e	1.911 ± 0.041 D
90	2.436 ± 0.060 f	2.274 ± 0.021 E *

3.2. *High-Pressure Homogenization for Enhanced Solubilization of Protein*

3.2.1. Effect of Number of HPH Passes

HPH can disrupt cells through high shear forces, turbulence, and cavitation to break microbial cells. The efficiency of cell disruption and the release of intracellular components depends on both the applied pressure and the number of HPH passes. In this study, the concentration of both the wet and spray-dried biomasses was adjusted to 50 g L^{-1} (5% dry matter) using water, and the effect of the number of HPH passes was investigated at an operating pressure of 35 k psi with the temperature maintained at 4 °C. It is evident from Figure 2 that the overall yields for both biomass types treated with HPH were significantly improved compared to alkaline extraction alone (Figure 1), thus highlighting the effectiveness of mechanical disruption in breaking the biomass to facilitate protein release. For the wet biomass, the solubilization yield plateaued at approximately 58%, with no significant differences observed between the first and subsequent passes. This indicates that the cell disruption of the wet biomass was highly effective after a single pass. On the other hand, spray-dried biomass exhibited a progressive increase in solubilization yield with the number of passes, rising from 18.5% after the first pass to around 41% after the fourth pass.

The lower yields for spray-dried biomass using HPH, compared to wet biomass, were likely caused by structural changes induced by high-pressure atomization and high temperature during the spray drying process [15,16]. The exposure to the high spray drying temperature of 190 °C might lead to protein denaturation and aggregation, reinforcing the cell envelope and reducing the protein solubility in water. Additionally, the lipid could undergo phase transition, reducing cell wall permeability. Some structural changes, such as membrane fluidity, can partially recover through hydration with water; however, protein denaturation and lipid phase transition may not be fully reversed. Hence, protein solubilization yield from the spray-dried biomass, even after four passes (41.2%), was lower than that achieved with wet biomass after the first pass (52.3%). Further studies are required to better understand the cellular changes induced by the spray drying process, which would improve the biorefinery process development. Since the difference in wet biomass protein solubilization yield between the first and second passes of HPH was only about 10% but required twice the energy, the single-pass HPH was chosen for subsequent process optimization.

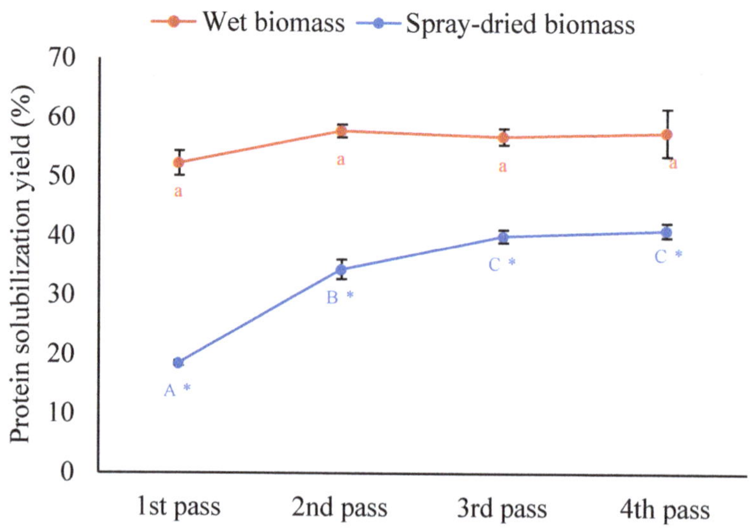

Figure 2. Effect of pass number on solubilization of wet and spray-dried *A. protothecoides* biomass during HPH with water. Different lowercase and uppercase letters indicate significant differences between treatments at different HPH passes for wet biomass and spray-dried biomass, respectively ($p < 0.05$). Asterisks (*) indicate significant differences between biomass types (wet vs. spray-dried) at the same HPH pass ($p < 0.05$).

3.2.2. HPH Followed by Alkaline Treatment for Enhanced Protein Solubilization

Following the first pass HPH with water, NaOH was added to adjust different alkaline concentrations, and the sample was mixed for 60 min for enhanced protein solubilization. An increasing trend in protein solubilization yield was observed for both wet and dry biomasses with the increase in NaOH concentration (Figure 3). The solubilization yield for dried biomass reached 68.0% at 0.1 M NaOH and remained almost the same at higher concentrations. In contrast, a similar yield of 68.1% was achieved with wet biomass at a 5-fold lower NaOH concentration of 0.02 M. Further increasing the NaOH concentration to 0.1 M led to a slight increase in yield to 70.3%. Notably, without NaOH treatment, about 2.8-fold more protein was released from the wet biomass (52.3%) compared to the dried biomass (18.5%) under the same conditions. These results suggest that the extraction of protein from wet biomass is more feasible in terms of enhanced solubilization yield, reduced energy consumption, and lower waste generation.

Figure 4 shows the impact of mixing time on the alkaline solubilization of HPH-treated wet biomass across different NaOH concentrations. It can be seen that a shorter mixing time of 5 min is sufficient for the solubilization of protein, and no considerable changes were observed with a longer mixing time. This result suggests that *A. protothecoides'* proteins after HPH-induced cell disruption could instantly be soluble in alkaline water, and the protein solubilization after HPH was mainly driven by alkaline concentration.

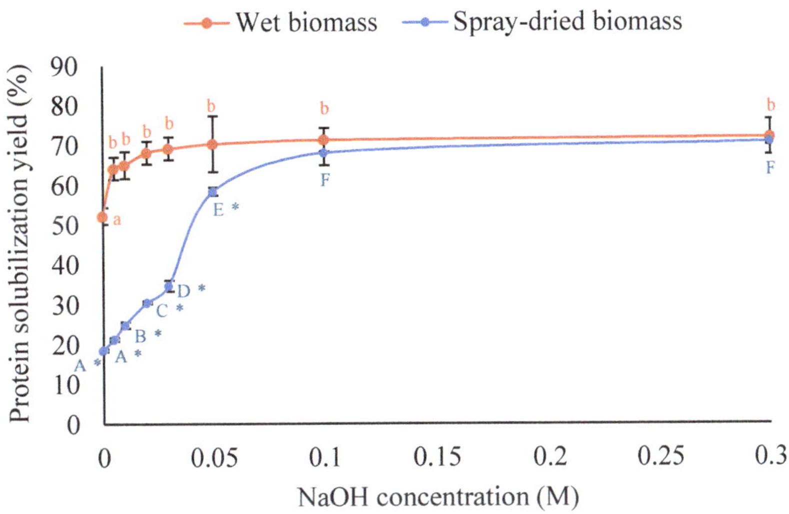

Figure 3. Effect of alkaline concentration on solubilization of first-pass homogenized wet and spray-dried *A. protothecoides* biomass. Different lowercase and uppercase letters indicate significant differences between treatments at different NaOH concentrations for wet biomass and spray-dried biomass, respectively ($p < 0.05$). Asterisks (*) indicate significant differences between biomass types (wet vs. spray-dried) at the same NaOH concentration ($p < 0.05$).

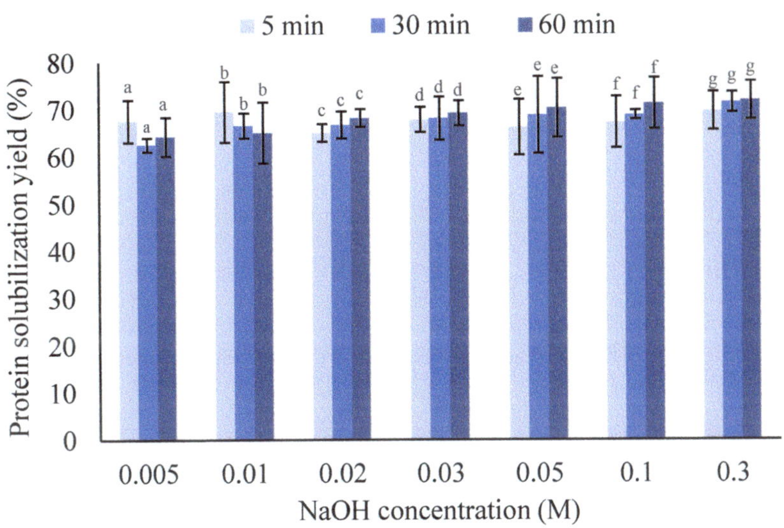

Figure 4. Effect of mixing time on protein solubilization yield of first-pass homogenized wet and spray-dried *A. protothecoides* biomass during alkaline solubilization step under different NaOH concentrations. Different lowercase letters indicate significant differences between treatments at different mixing times of the same NaOH concentration ($p < 0.05$). Asterisks (*) indicate significant differences between NaOH concentrations at the same mixing time ($p < 0.05$).

3.3. *Effect of Bead Milling on Protein Solubilization*

Because HPH induces cell rupture through a combination of sudden pressure drop, shear force, turbulence, and cavitation, it is usually preferred for soft- to moderate-walled cells. In contrast, bead milling, where agitation causes the collision between beads and cells to physically break them, may be more effective for tough-walled cells. Hence, bead milling followed by alkaline treatment was applied to wet and spray-dried *Auxenochlorella* biomass for enhanced protein extraction. It is evident from Figure 5 that protein solubilization yields for both bead-milled wet and spray-dried biomasses increased with increasing NaOH concentrations. The maximum solubilization yield with wet biomass reached 70.5% at 0.05 M NaOH, about 2.5-fold higher than that achieved with dried biomass under the same conditions. This result further confirms that proteins from wet biomass are readily released into alkaline solution compared to those in spray-dried biomass, and that cell disruption is essential prior to alkaline treatment for enhanced protein extraction.

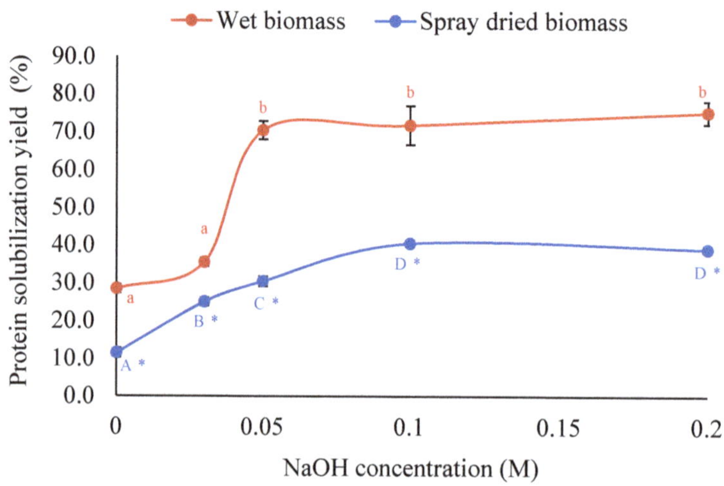

Figure 5. Effect of NaOH concentration on protein solubilization of bead-milled wet and spray-dried *A. protothecoides* biomass during alkaline solubilization step. Different lowercase and uppercase letters indicate significant differences between treatments at different NaOH concentrations for wet biomass and spray-dried biomass, respectively ($p < 0.05$). Asterisks (*) indicate significant differences between biomass types (wet vs. spray-dried) at the same NaOH concentration ($p < 0.05$).

3.4. Comparison Between HPH and Bead Milling for Protein Solubilization

The effectiveness of HPH (35 k psi, single pass) and bead milling (26 Hz, 60 min) was compared in the solubilization of protein from wet biomass at different NaOH concentrations (Figure 6). Both pre-treatment methods improved protein solubilization. For bead-milled biomass, alkaline treatment had minimal impact at a lower NaOH (<0.03 M) concentration, while a sharp increase in solubilization yield was observed between 0.03 and 0.05 M NaOH. The maximum yield reached 70.5% after which it plateaued. In contrast, the HPH process yielded a comparable yield at a lower NaOH concentration of 0.03 M, demonstrating superior efficacy over bead milling. It is worth noting that the HPH process yielded 1.8-fold higher protein solubilization (52.3%) without any NaOH (at 0.0 M) compared to bead milling (28.5%). These results suggest that HPH is preferred to disrupt *A. protothecoides* microalgal cells for enhanced protein extraction.

Alternative cell disruption methods have also been utilized to extract protein from *A. protothecoides*. Perez et al. [8] reported improved protein extraction yields by combining PEF or ultrasonication (US) with a 24 h incubation after treatment, resulting in a 122% increase in yield for PEF (from 11% to 25%) and 51.9% for US (from 13% to 19%). While PEF and US could be more energy efficient compared to HPH, the longer incubation time of 24 h makes the overall process slower. However, the proposed HPH method shows an advantage of being more effective, yielding higher protein recovery up to 70% without the need for extended incubation, which provides a more time-efficient and potentially more scalable solution for protein extraction from wet microalgal biomass. Further investigation is necessary to evaluate the energy efficiency of the proposed method. Additionally, a thorough assessment of protein quality and functional properties remains important regardless of PEF, US, and HPH. Mechanical cell disruption methods have been shown to significantly enhance protein digestibility without markedly altering amino acid profiles [17]. Moreover, the effects of cell disruption on sensory characteristics should be examined, as they can greatly impact the flavor profile and consumer acceptability of microalgae-based food products. For example, it was reported that HPH processing of *Nannochloropsis* generates intense grassy and fishy odors caused by elevated levels of fatty acid-derived unsaturated aldehydes, ketones, and alcohols [18].

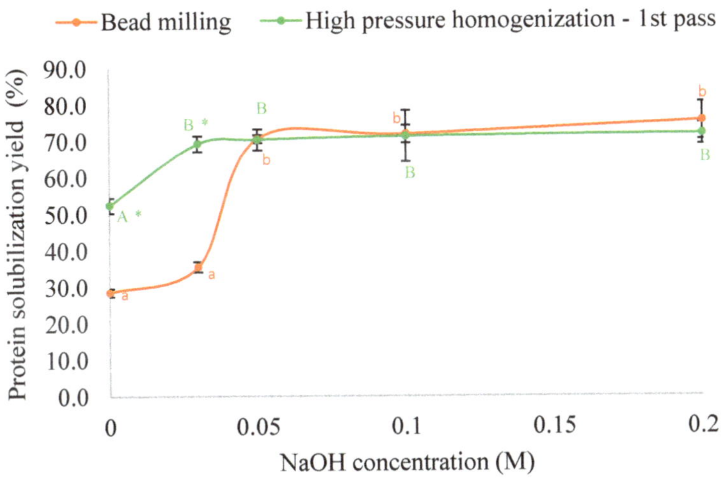

Figure 6. Comparison of protein solubilization yield between bead-milled and first-pass homogenized wet *A. protothecoides* biomass during the alkaline solubilization step. Different lowercase and uppercase letters indicate significant differences between treatments at different NaOH concentrations for the bead-milling method and high-pressure homogenization method, respectively ($p < 0.05$). Asterisks (*) indicate significant differences between the bead milling and high-pressure homogenization methods at the same NaOH concentration ($p < 0.05$).

4. Conclusions

A method that eliminates energy-consuming microalgae biomass drying and significantly reduces the requirement of NaOH concentration was developed for effective extraction of microalgae protein. HPH alone without alkaline treatment released 52.3% of protein from wet microalgae biomass, which is 2.8-fold higher compared to that achieved using spray-dried biomass (18.5%). HPH with subsequent mild alkaline treatment further increased the protein extraction yield from 52.3 to 68.1% with 0.02 M NaOH. The protein extraction yield further increased to 71.2% using a higher NaOH concentration. However, high NaOH concentrations could adversely affect the protein attributes for food applications. The reduced alkaline loading is particularly important in large-scale production, where the environmental impact of chemicals and waste disposal is a crucial consideration. Furthermore, it could potentially contribute to higher-quality protein extracts with more promising techno-functional properties.

Additionally, HPH proved superior to bead milling for disrupting microalgae cell walls, enabling protein extraction at a reduced NaOH concentration. The maximum protein extraction yield with bead milling was 70.5% using 0.05 M NaOH, whereas HPH achieved the same yield with only 0.03 M NaOH. Notably, HPH alone without alkaline treatment led to a 1.8-fold protein solubilization compared to bead milling.

Future research should focus on investigating the techno-functional properties, including the quality, such as the amino acid composition, digestibility, and bioactivity of the protein extracts obtained using this optimized method. These properties are critical for the use of microalgae proteins in food and nutraceutical applications.

Author Contributions: Conceptualization, J.W.N. and M.M.R.T.; methodology, J.W.N., T.M.T. and M.M.R.T.; software, J.W.N.; validation, J.W.N., S.Y.L. and M.M.R.T.; formal analysis, J.W.N. and S.Y.L.; investigation, J.W.N. and S.Y.L.; resources, J.W.N., S.Y.L. and M.M.R.T.; data curation, J.W.N. and S.Y.L.; writing—original draft preparation, J.W.N. and S.Y.L.; writing—review and editing, M.M.R.T. and M.W.; visualization, J.W.N. and S.Y.L.; supervision, M.M.R.T.; project administration, M.M.R.T. and M.W.; funding acquisition, M.M.R.T. and M.W. All authors have read and agreed to the published version of the manuscript.

Funding: This research was funded by the Singapore National Research Foundation INTRA-CREATE THEMATIC GRANT (NRF2020-THE003-0004) and Agency for Science, Technology and Research (A*STAR) as part of the Singapore Food Story R&D Programme (Grant number A21H7a0132).

Data Availability Statement: The original contributions presented in the study are included in the article, further inquiries can be directed to the corresponding author.

Conflicts of Interest: The authors declare no conflicts of interest.

References

1. Colla, L.M.; Oliveira Reinehr, C.; Reichert, C.; Costa, J.A.V. Production of Biomass and Nutraceutical Compounds by *Spirulina platensis* under Different Temperature and Nitrogen Regimes. *Bioresour. Technol.* **2007**, *98*, 1489–1493. [CrossRef] [PubMed]
2. Morris, H.J.; Almarales, A.; Carrillo, O.; Bermúdez, R.C. Utilisation of *Chlorellavulgaris* Cell Biomass for the Production of Enzymatic Protein Hydrolysates. *Bioresour. Technol.* **2008**, *99*, 7723–7729. [CrossRef] [PubMed]
3. Becker, E.W. Micro-Algae as a Source of Protein. *Biotechnol. Adv.* **2007**, *25*, 207–210. [CrossRef] [PubMed]
4. Ng, J.W.; Teh, T.M.; Tan, C.F.; Bi, X.; Low, Z.E.; Rahman Talukder, M.M. Alkaline Solubilization of Microalgal Protein and Its Impact on the Functional Properties of Protein Extract. *Future Foods* **2024**, *9*, 100368. [CrossRef]
5. Safi, C.; Frances, C.; Ursu, A.V.; Laroche, C.; Pouzet, C.; Vaca-Garcia, C.; Pontalier, P.-Y. Understanding the Effect of Cell Disruption Methods on the Diffusion of *Chlorella vulgaris* Proteins and Pigments in the Aqueous Phase. *Algal Res.* **2015**, *8*, 61–68. [CrossRef]
6. Postma, P.R.; Miron, T.L.; Olivieri, G.; Barbosa, M.J.; Wijffels, R.H.; Eppink, M.H.M. Mild Disintegration of the Green Microalgae *Chlorella vulgaris* Using Bead Milling. *Bioresour. Technol.* **2015**, *184*, 297–304. [CrossRef] [PubMed]
7. Jubeau, S.; Marchal, L.; Pruvost, J.; Jaouen, P.; Legrand, J.; Fleurence, J. High Pressure Disruption: A Two-Step Treatment for Selective Extraction of Intracellular Components from the Microalga *Porphyridium cruentum*. *J. Appl. Phycol.* **2012**, *25*, 983–989. [CrossRef]
8. Perez, B.; Zermatten, C.; Haberkorn, I.; Mathys, A. Enhancing protein extraction from heterotrophic *Auxenochlorella protothecoides* microalgae through emerging cell disruption technologies combined with incubation. *Bioresour. Technol.* **2024**, *407*, 131099. [CrossRef] [PubMed]
9. Katsimichas, A.; Karveli, I.; Dimopoulos, G.; Giannakourou, M.; Taoukis, P. Kinetics of High Pressure Homogenization Assisted Protein Extraction from *Chlorella pyrenoidosa*. *Innov. Food Sci. Emerg. Technol.* **2023**, *88*, 103438. [CrossRef]
10. Chen, Y.; Chen, J.; Chang, C.; Chen, J.; Cao, F.; Zhao, J.; Zheng, Y.; Zhu, J. Physicochemical and Functional Properties of Proteins Extracted from Three Microalgal Species. *Food Hydrocoll.* **2019**, *96*, 510–517. [CrossRef]
11. Ng, J.W.; Teh, T.M.; Weingarten, M.; Rahman Talukder, M.M. Integrating Bead Milling and Alkaline Solubilization for Enhanced Protein Recovery from Microalgae: A Comprehensive Approach. *Future Foods* **2024**, *9*, 100385. [CrossRef]
12. Bradford, M.M. A Rapid and Sensitive Method for the Quantitation of Microgram Quantities of Protein Utilizing the Principle of Protein-Dye Binding. *Anal. Biochem.* **1976**, *72*, 248–254. [CrossRef] [PubMed]
13. Chia, S.R.; Chew, K.W.; Zaid, H.F.M.; Chu, D.-T.; Tao, Y.; Show, P.L. Microalgal Protein Extraction from *Chlorella vulgaris* FSP-E Using Triphasic Partitioning Technique with Sonication. *Front. Bioeng. Biotechnol.* **2019**, *7*, 396. [CrossRef] [PubMed]
14. Rashidi, B.; Trindade, L.M. Detailed Biochemical and Morphologic Characteristics of the Green Microalga *Neochloris oleoabundans* Cell Wall. *Algal Res.* **2018**, *35*, 152–159. [CrossRef]
15. Alhattab, M.; Kermanshahi-Pour, A.; Brooks, M.S.-L. Microalgae Disruption Techniques for Product Recovery: Influence of Cell Wall Composition. *J. Appl. Phycol.* **2018**, *31*, 61–88. [CrossRef]

16. Lin, L.P. Microstructure of Spray-Dried and Freeze-Dried Microalgal Powders. *Food Struct.* **1985**, *4*, 17.
17. Wang, Y.; Tibbetts, S.M.; Berrue, F.; McGinn, P.J.; MacQuarrie, S.P.; Puttaswamy, A.; Patelakis, S.; Schmidt, D.; Melanson, R.; MacKenzie, S.E. A Rat Study to Evaluate the Protein Quality of Three Green Microalgal Species and the Impact of Mechanical Cell Wall Disruption. *Foods* **2020**, *9*, 1531. [CrossRef] [PubMed]
18. Coleman, B.; Van Poucke, C.; De Witte, B.; Casciaro, V.; Moerdijk-Poortvliet, T.; Muylaert, K.; Robbens, J. The effect of drying, cell disruption and storage on the sensory properties of *Nannochloropsis* sp. *Algal Res.* **2023**, *71*, 103092. [CrossRef]

foods

MDPI

Article

Emulsification Properties of Plant and Milk Protein Concentrate Blends

Mohammadreza Khalesi [1,2,3,*], Shauna Dowling [1], Jack Comerford [1], Ciara Sweeney [1], Sara Esteghlal [4] and Richard J. FitzGerald [1,2,3]

1 Department of Biological Sciences, University of Limerick, V94 T9PX Limerick, Ireland; ciarasweeney1010@gmail.com (C.S.)
2 Bernal Institute, University of Limerick, V94 T9PX Limerick, Ireland
3 Health Research Institute, University of Limerick, V94 T9PX Limerick, Ireland
4 Department of Food Science and Technology, Shiraz University, Shiraz 71441-65186, Iran; sara.esteghlal@shirazu.ac.ir
* Correspondence: mohammadreza.khalesi@ul.ie

Abstract

Blending is a promising strategy during the partial replacement of plant with animal proteins. This, however, may lead to alteration in the technofunctional properties of the resultant blends. In this study, partial replacement of milk protein concentrate (MPC) with different plant proteins including soy, rice and pea protein concentrates (SPC, RPC and PPC, respectively) was conducted to determine the effect of blending at different ratios on the technofunctional properties relevant to their emulsification behavior, e.g., emulsion stability, viscosity and water holding capacity (WHC) and oil binding capacity (OBC). It was observed that at equivalent concentrations, the plant protein concentrates had higher apparent viscosities compared to MPC and the blends. RPC–MPC, at all ratios (25:75, 50:50, and 75:25), had a lower OBC when compared with the SPC–MPC and PPC–MPC blends. The lowest OBC was 32.5, for RCP–MPC 25:75, and the highest was 116.0 for SPC–MPC 25:75. The highest solubility of PPC, RPC, and SPC was observed in their blend form at 50:50 (73.2%), 75:25 (86.5%) and 25:75 (71.1%) ratios, respectively. Plant protein–MPC blends showed higher emulsion stability than the individual plant protein concentrates. The highest emulsion stability was 100%, for RPC–MPC 50:50 and 75:25 ratios, PPC–MPC at 50:50 ratio, and SPC–MPC at 25:75 and 100:0 ratios. Among the blends, SPC–MPC 25:75, PPC–MPC 50:50 and RPC–MPC 50:50 showed the most suitable overall emulsification properties. Based on the results, blending MPC with plant protein concentrates led to promising improvements in emulsification behavior relevant to different composite protein ingredient applications.

Keywords: plant protein; milk protein concentrate; blend; emulsion properties

1. Introduction

The global population is progressively increasing, leading to a growing demand for dietary protein [1]. Currently, the proteins used to meet these demands mainly originate from animal sources, e.g., meat and dairy proteins. Milk protein concentrates (MPCs) are dairy ingredients containing 42–85% protein [2]. MPCs are used in products such as in dietary supplements, nutrition bars and sports beverages due to their high level of essential amino acids (EAAs) along with their low lactose content, particularly in the case of higher-protein-content MPCs.

While animal proteins have numerous advantages in food product applications, the latest developments in food sustainability focus on the impact of animal-origin products, particularly in terms of their climate change and economic implications. Accordingly, the demand for alternative protein sources, mainly from plants, is increasing. Many plant proteins have good functional (e.g., emulsification) properties in food applications, which enables them to serve as potential substitutes for animal-origin proteins. Nevertheless, there are some issues that may restrict the widespread application of plant proteins in food products. For instance, some plant proteins contain low levels of certain EAAs and high levels of anti-nutritional factors [3]. In addition, some plant proteins display relatively poor technofunctional properties [4].

Therefore, there is an increased interest in the functionality and nutritional quality of hybrid protein products, i.e., combinations of plant with animal proteins [5,6]. Emerging evidence suggests that blending plant with animal proteins could increase the utilization of plant proteins while improving their overall functionality and nutritional quality. Blending can also be considered a new marketing opportunity for food manufacturers to develop products with novel characteristics (such as improved technofunctionality and sensory properties).

Alves and Tavares [7] stated that the partial replacement of animal protein with plant protein is the first step towards reducing the environmental impacts associated with animal food consumption. There are different studies exploring the emulsification properties of plant protein and animal protein blends. Blending of pea protein isolate (PPI) and whey protein isolate (WPI) improved the functionality of PPI in a mixed-protein system formulation [8]. It has also been shown that blending of skim milk powder with pea protein concentrate (PPC) modified the technofunctional properties of first age infant formula [9]. Blending increased the viscosity and reduced the solubility, but did not change the emulsion stability (ES) of the product. Ho et al. [10] reported that plant-derived emulsifiers generated with soy protein isolate (SPI) and PPI were suitable replacements for dairy proteins including WPI and sodium caseinate. It has been reported that milk protein–soy protein (SP) blends have higher apparent viscosity (η_{app}) compared to micellar casein (CN) [11]. For example, potato protein (PP) and whey protein (WP) were blended at different ratios and then used to stabilize oil in water emulsions. The results indicated that by increasing PP content, the viscosity of the emulsions, the amount of adsorbed protein at the oil–water interface, surface protein load, and thickness of the interfacial film were increased, which resulted in emulsions with higher stability. The ratio of 7:3 PP to WP was indicated as the optimal ratio in terms of emulsification aspects [12]. In another study, WPI, SPI, and their blending at 1:1 ratio were used to stabilize emulsions with 5, 10, and 15% oil concentrations. The blended proteins revealed a higher protein adsorption rate than the WPI, and lower than that of SPI. Higher oil concentrations in the emulsions led to an increase in protein adsorption (especially the blend and SPI). The blended proteins resulted in an emulsion with the highest stability [13]. Blending SPI and WPI at different ratios was observed to efficiently decrease the size of emulsion droplets, increase the thickness of the interfacial layer, and enhance emulsion stability. The ratio of 1:9 of SPI to WPI and protein concentration of 0.2 g/kg led to the lowest creaming index and smallest droplet size. The highest protein adsorption to the interfacial layer was observed at a protein concentration of 0.15 g/kg and SPI to WPI mass ratio of 9:1 [14]. Partial replacement of WPI or sodium caseinate with PPI in stabilized emulsions showed that the blend of proteins exhibited higher surface load and droplet size compared to the individual animal proteins. Blending the plant and animal proteins resulted in enhanced storage stability of the emulsions as a result of synergistic effects [15].

Among different plant proteins, soy protein is one of the primary candidates for partially replacing animal proteins. This protein is commercially present in the market and has broad applications owing to its high protein concentration, appropriate cost, and high consumer acceptance among people. Pea protein and rice protein are other promising plant protein sources for the partial replacement of animal proteins as they have desirable properties such as hypoallergenic character with a good balance of amino acids, significant digestibility, high biological value, and anti-cancer and anti-oxidant activities. Rice protein has a nutritional quality similar to or even better than other cereals. In addition to containing 20% proteins with higher lysine and tryptophan content compared to cereal grains, pea protein has 5–20% lower trypsin inhibitors than soybeans. [16,17].

Despite the numerous reports on blending plant and animal proteins, it is still relatively unknown how plant and dairy proteins may behave when in blends. Limited knowledge appears to exist on the impact of blending of dairy and plant proteins on the functionality, e.g., emulsification properties, in different products. Therefore, acquisition of this knowledge may help in the targeted design of balanced blends for different functional and nutritional applications.

The hypothesis is that blending plant with animal proteins has the potential to yield protein mixtures with novel emulsion properties due to the potential for interactive effects between the different origin proteins. This in turn may lead to the development of new functionality and ingredient applications. The objective of this study was to evaluate the impact of blending different plant proteins, i.e., soy protein concentrate (SPC), rice protein concentrate (RPC), and PPC with MPC at different ratios on properties relevant to the emulsification behavior of blends. In addition, the optimal plant proteins and blending ratios in terms of improved emulsification properties will be introduced. To our knowledge, this appears to be the first report on the emulsion properties of plant protein–MPC blends.

2. Materials and Methods

2.1. Materials

SPC, PPC and RPC from Pulsin Ltd. (Gloucester, UK) were obtained at a local health food store and MPC85 (85% (*w*/*w*) protein) was obtained from a commercial manufacturer. Corn oil was purchased from a local food market. Sodium hydroxide (NaOH) and acetic acid were from Fisher Scientific (Dublin, Ireland). Kjeldahl catalyst tablets, sulfuric acid (>98%), boric acid, 2-mercaptoethanol, methanol, protein molecular mass markers (6.5–200.0 kDa) and Sudan III were from Sigma-Aldrich (Dublin, Ireland). Hexane was from Honeywell International Inc. (Dublin, Ireland). Coomassie R, Laemmli buffer, Mini-Protean TGX 4–20% pre-cast polyacrylamide gels were from Bio-Rad Laboratories Inc. (Hercules, CA, USA) and sodium dodecyl sulfate (SDS) was from National Diagnostics (Atlanta, GA, USA).

2.2. Proximate Analysis and pH Determination

Moisture, ash, lipid and protein contents were determined (n = 3) according to Khalesi and FitzGerald [18].

2.3. Blending of Plant Protein Samples with MPC

Different blends having different ratios of proteins from plant sources and MPC85 were generated as schematically outlined in Figure 1. In addition, the quantities of each protein at each ratio are shown in Table 1.

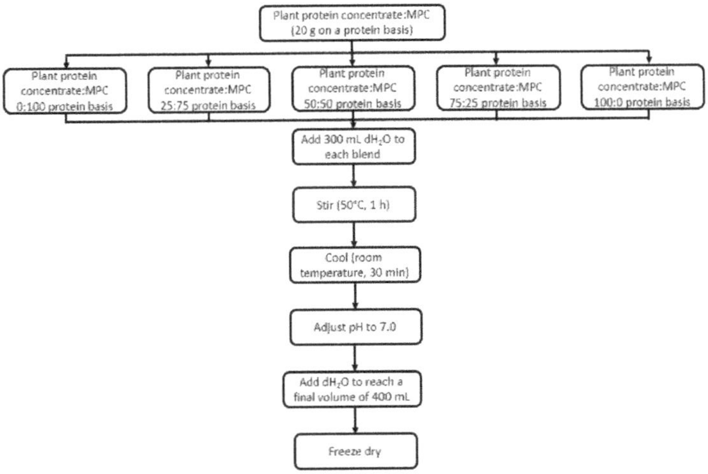

Figure 1. Schematic of the process of preparing blends of the plant protein concentrates (i.e., soy (SPC), pea (PPC) and rice (RPC) protein concentrates) and milk protein concentrate (MPC).

Table 1. Quantities of each protein type at each ratio.

Protein Type	Plant Protein to Milk Protein Ratio				
	100:0	75:25	50:50	25:75	0:100
Plant protein (g)	20	15	10	5	0
Milk protein (g)	0	5	10	15	20
Total weight of blend (g)	20	20	20	20	20

2.4. *Technofunctional Property Analysis*

2.4.1. Emulsification

Freeze-dried samples of each individual plant protein concentrate, MPC and the plant protein–MPC blends were resuspended with dH_2O and adjusted to pH 7.0 to give a 0.025% (*w*/*v*) protein suspension. Sudan Red III (40 mg) was added to 1 L of corn oil, after which 6 g was added to 14 g of each protein sample suspension. Samples were then homogenized using an Ultra-Turrax (IKA T25, Staufen, Germany) for 1 min at 16,000 rpm in order to create an emulsion. Immediately after homogenization, an aliquot of sample (18 μL) was 100 fold diluted with 0.1% (*w*/*v*) SDS to reach the volume of 1.8 mL. The absorbance (A, λ500) of the bottom half of the emulsion sample was measured (n = 3) using a UV–Vis 1800 spectrophotometer (Shimadzu, Canby, OR, USA) at T0 and 30 min (T30) after emulsion formation. ES was determined according to Equation (1):

$$ES(\%) = \frac{A_{T30}}{A_{T0}} \times 100 \tag{1}$$

where AT30 and AT0 represent the absorbance (λ500) at T30 and T0 (min), respectively.

2.4.2. Apparent Viscosity (η_{app})

An aliquot (16 mL) of each suspension equivalent to 5% (*w*/*v*) protein prepared after reconstitution of the freeze-dried samples was analyzed using a Brookfield DV-II viscometer (Analytica, Dublin, Ireland) at 30 °C (n = 3) at a shear rate of 6 s^{-1} for the PPC, SPC and their MPC blends and at 100 s^{-1} for the RPC-containing samples. The η_{app} of MPC was measured at both share rates (i.e., 6 and 100 s^{-1}).

2.4.3. Water Holding Capacity (WHC) and Oil Binding Capacity (OBC)

The WBC/OBC for the different freeze-dried blends was determined (n = 3) by re-suspension of each sample/blend in dH_2O/corn oil to reach a final concentration of 5% (w/v) on a protein basis, vortexed for 30 s followed by centrifugation (320R Hettich centrifuge, Tuttlingen, Germany) at 3000 g at 20 °C for 30 min and then, the supernatant (water/oil) was removed. The WHC (g dH_2O/g protein)/OBC (g oil/g protein) was calculated according to Equation (2) [18].

$$WHC/OBC(\%) = \frac{\text{W3} - \text{W1}}{\text{W2}} \times 100 \quad (2)$$

where W1, W_2 and W_3 are the mass of the freeze-dried sample, weight of protein in each powder sample (as determined in Section 2.2), and the weight of residue after centrifugation followed by removal of the supernatant, respectively.

2.4.4. Solubility

To determine the solubility of the proteins/blends (n = 3), certain amount of each sample was dispersed in dH_2O to reach 5% (*w/v, protein basis*) concentration and then stirred for 30 min at RT to ensure complete hydration. The dispersion was then centrifuged (320R Hettich centrifuge, Tuttlingen, Germany) at 1000 g for 10 min at 25 °C. The supernatant was carefully collected and its total solids content was determined after drying at 95 °C for 6 h in a vacuum oven (OVA03100, Gallenkamp Ltd., Loughborough, UK). Solubility of the samples was then calculated based on the below eq [18].

$$Solubility(\%) = \frac{\text{Total solids of supernatant}}{\text{Total solids of protein dispersion}} \times 100 \quad (3)$$

2.5. Statistical Analysis

Each test was conducted three times (n = 3). Data values were presented as the mean ± standard deviation (SD). One-way analysis of variance (ANOVA) followed by the Tukey post hoc comparison test was carried out to test for significant differences using Minitab® Release 15 for Windows. A p value < 0.05 was considered statistically significant.

3. Results and Discussion

3.1. Proximate Analysis and Sample Properties

As shown in Table 2, it is evident that there was some variation in the moisture, protein, ash and the lipid contents of the different test samples. The moisture content in the samples ranged from 1.64 to 5.58% with the lowest value being for RPC and the highest value for SPC. Similarly to our results, Kaspchak et al. reported that among soy, pea, and rice protein isolates, the lowest and highest moisture content were for rice protein isolate (1.43%) and soy protein isolate (3.04%), respectively [19]. MPC had the highest overall protein content (84.17 ± 0.79%), while among the plant protein concentrates, SPC had the highest protein content (81.11 ± 0.77%). The protein content obtained for PPC was 71.01 ± 0.25%. Elsewhere, the chemical analysis of pea, soy, and rice proteins indicated that the protein content was in the range of 76.96–86.36%, with the highest protein content for soybean protein [17]. Similarly, proximate analysis of the same proteins in another study indicated the amount of 89.99%, 88.00%, and 82.76% protein for soy, pea, and rice protein isolates, respectively [19]. The lipid content of MPC85 (1.31 ± 0.07%) was the lowest among the samples as it is manufactured from skim milk. Significant differences were found between the lipid content in each of the plant protein samples. The lipid content in RPC (9.70 ± 0.37%) and PPC (8.13 ± 0.17%) was higher than in SPC (1.79 ± 0.11%).

The ash content showed less variance, the highest mean value being 6.96% in MPC and the lowest being 5.47% for SPC. PPC had the highest mean ash level (6.48%) among the plant protein samples. The amount of measured ash by Zhao et al. in soy, rice, and pea protein samples was reported to have lower values, 0.24, 3.83, and 4.01 for rice, pea, and soy proteins, respectively [17]. The lower ash content may be attributed to different varieties of the plants and the protein extraction method.

Table 2. Moisture, protein, ash and lipid content along with reconstitution pH (at 0.05% (*w*/*v*) on a protein basis) of soy (SPC), rice (RPC) and pea protein concentrates (PPC) and milk protein concentrate (MPC).

Sample	Moisture	Protein	Ash	Lipid	pH
			(%)		
SPC	5.58 ± 1.57 [a]	81.11 ± 0.77 [a]	5.47 ± 0.27 [a]	1.79 ± 0.11 [a]	7.15 ± 0.06 [a]
RPC	1.64 ± 0.41 [b]	80.04 ± 0.88 [a]	5.50 ± 0.29 [a]	9.70 ± 0.37 [b]	6.09 ± 0.04 [b]
PPC	3.86 ± 0.01 [a]	71.01 ± 0.25 [b]	6.48 ± 0.47 [b]	8.13 ± 0.17 [c]	8.00 ± 0.06 [c]
MPC	4.82 ± 0.02 [a]	84.17 ± 0.79 [c]	6.96 ± 0.16 [b]	1.31 ± 0.07 [d]	7.09 ± 0.03 [a]

Values represent the mean ± SD (n = 3). Different letters represent significant differences within each column ($p < 0.05$).

The reconstitution pH values recorded ranged from pH 6 to 8, with RPC being slightly acidic (pH 6.09) and PPC being slightly basic (pH 8.00). The mean pH of SPC was similar to MPC (7.15 vs. 7.09).

3.2. Technofunctional Properties

3.2.1. Apparent Viscosity (η_{app})

The η_{app} varied across the plant blends (Figure 2). Overall, SPC and PPC produced the most viscous suspensions. The η_{app} for SPC (64.3 ± 10.9 mPa.s) and PPC (62.5 ± 8.7 mPa.s) were not significantly different ($p > 0.05$). Both 100% SPC and PPC gave a higher viscosity value than MPC (36.8 ± 1.2 mPa.s) at a similar shear rate. Previously, the η_{app} of SPC was reported to be similar to sodium caseinate but was two times higher when compared to WPI [20]. The results herein demonstrated that PPC and SPC have, under certain circumstances, elevated viscosity properties which may be desirable for emulsions as well as in the formulation of high viscosity requiring products, e.g., plant-based yogurts and ice cream. The η_{app} of RPC (1.4 ± 0.1 mPa.s) was not significantly ($p > 0.05$) different from MPC (1.5 ± 0.1 mPa.s) when tested at a similar shear rate (i.e., 6 s^{-1}). The mean viscosity values of the SPC–MPC blends tended to increase as the proportion of SPC in the blends increased. This is associated with the higher η_{app} of SPC compared to MPC. The η_{app} associated with the blends generated with PPC was higher ($p < 0.05$) than those generated with SPC, except when at a ratio of PPC–MPC of 50:50 (Figure 2a). A previous study showed that the inclusion of PPC in infant formula produced with skim milk (50:50) enhanced the viscosity while the inclusion of faba bean protein at the same ratio did not change the overall viscosity [21]. The η_{app} of WPI was reported to be significantly increased on blending with PPI [22]. The η_{app} of sodium caseinate was reported to be higher than PPI and a PPI-sodium caseinate hybrid blend [23]. The reconstituted suspensions of the RPC–MPC blends had the lowest η_{app} among the plant–MPC blends (Figure 2b). Increasing the quantity of RPC did not increase the η_{app}, with the RPC–MPC 25:75 blend yielding the highest η_{app} value (3.76 ± 0.13 mPa.s) among the RPC–MPC blends ($p < 0.05$). The results showed that the hybrid blends created with RPC and MPC had a higher η_{app} compared to MPC and RPC alone, indicating possible interactions between the RPC and MPC protein suspensions.

Overall, the PPC–MPC and SPC–MPC blends may prove useful for high viscosity requiring applications. It should also be noted that the presence of other non-proteinaceous components may cause differences between the η_{app} of various plant protein ingredients. In addition, most of the blend samples gave η_{app} values higher than that for MPC alone, suggesting the possibility of replacement of MPC with plant proteins for high viscosity requiring purposes. In general, higher viscosity is associated with better emulsification properties as in a more viscose medium, the mobility of the oil droplets is reduced, which results in lower coalescence within the emulsion [24].

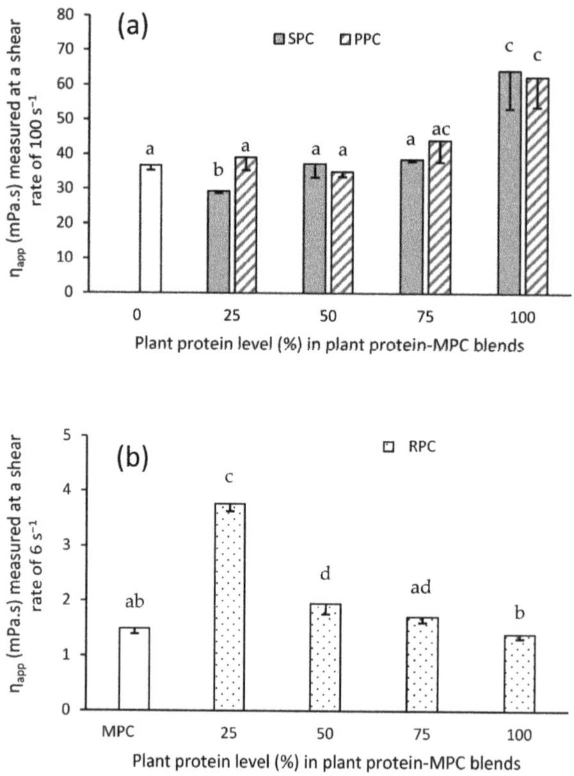

Figure 2. The apparent viscosity (η_{app}) (**a**) of soy (SPC) and pea (PPC) protein concentrates blended with milk protein concentrate (MPC) at different ratios, i.e., 0:100, 25:75, 50:50, 75:25 and 100:0, at a shear rate of 6 s^{-1}, and (**b**) of rice protein concentrate (RPC) blended with MPC at different ratios, i.e., 0:100, 25:75, 50:50, 75:25 and 100:0, at a shear rate of 100 s^{-1}. Values represent the mean ± SD (n = 3). Different letters represent significant differences ($p < 0.05$).

3.2.2. WHC and OBC

MPC had a higher ($p < 0.05$) WHC (349 ± 24 g water/100 g protein) compared to the individual plant protein concentrates tested (Figure 3). The extensive interaction between protein components, especially the CNs, in MPC and water molecules is considered the main reason for the high WHC of MPC. In addition, the plant protein concentrates had a higher lipid content, thus expectedly; they showed a lower affinity to retain water compared to MPC. SPC had a higher ($p < 0.05$) WHC (184 ± 8 g water/100 g protein) than RPC (26 ± 2 g water/100 g protein) and PPC (129 ± 5 g water/100 g protein). Some specific protein components of SPC (particularly the 11S globulin) have previously been shown to contribute to the WHC and to the formation of stable protein gels [25]. The higher WHC of SPC compared to PPC and RPC may also be associated with the lower lipid content of the SPC ingredient studied herein. Among the blended samples, SPC–MPC 25:75

(394 ± 9 g water/100 g protein) had the highest ($p < 0.05$) WHC. As the proportion of SPC increased and MPC decreased in the SPC–MPC blends, the WHC decreased. All SPC–MPC blends had a notably higher WHC in comparison to the RPC–MPC and PPC–MPC blends. Minimal variation was observed in the WHC between any of the RPC–MPC blends where the RPC–MPC blends had the lowest WHC (23–33 g water/100 g protein). The WHC of the PPC–MPC blends was significantly lower ($p < 0.05$) than the MPC sample. The WHC of the PPC–MPC 25:75 blend (129 ± 5 g water/100 g protein) was higher than 100% PPC (94 ± 2 g water/100 g protein) ($p < 0.05$). These results indicate interactions between SPC–MPC (25:75) resulting in an improvement in the WHC. This may be beneficial for the partial replacement of MPC for applications where high WHC and gelation properties are required, e.g., in yogurt and cheese-type products.

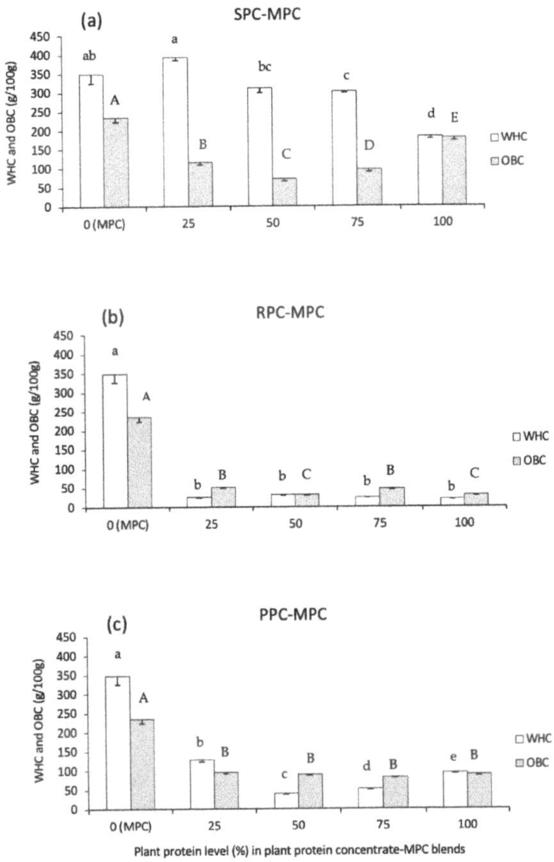

Figure 3. Water holding (WHC) and oil binding capacity (OBC) of (**a**) soy (SPC), (**b**) rice (RPC) and (**c**) pea protein concentrates (PPCs) blended with milk protein concentrate (MPC) at different ratios, i.e., 0:100, 25:75, 50:50, 75:25 and 100:0. Values represent the mean ± SD ($n = 3$). Different lower case letters represent significant differences ($p < 0.05$) for WHC values, and different upper case letters represent significant differences ($p < 0.05$) for OBC values.

Variability in the results of OBC of individual plant proteins was observed which may be associated with the composition of the plant proteins, especially the lipid and protein contents and also the surface located composition of the powder particles. As shown in Figure 3, no plant protein concentrates or plant protein–MPC blend reached an

OBC similar to that of MPC (239 g oil/g protein). This may be related to the presence of high levels of surface lipid in MPC85 which has been shown to consist of >15% of its surface composition [26]. In addition, the low OBC of plant proteins has previously been reported to be due to a large proportion of hydrophilic protein groups on their surfaces [27]. The OBC of SPC (180 ± 8 g oil/100 g protein) and SPC–MPC blends (ranging between 66 and 123 g oil/100 g protein) was highest amongst the plant protein concentrates and plant protein–MPC blends. RPC and RPC–MPC blends performed the poorest across all plant protein concentrates and plant protein–MPC blends in terms of OBC. Among the RPC–MPC blends, RPC–MPC 25:75 yielded the highest OBC at 51 ± 2 g oil/100 g protein, while the lowest OBC (33 ± 2 g oil/100 g protein) was seen for RPC–MPC 50:50. The OBC of PPC was 89 ± 3 g oil/100 g protein, which is in the range of previous reports showing that the OBC of commercial PPC was ~100 g oil/100 g protein. The ratio of 11S to 7S globulins was suggested to have an impact on the OBC of PPI [28]. Among PPC–MPC blends, PPC–MPC 25:75 yielded the highest OBC at 96 ± 4 g oil/100 g protein, while the lowest OBC (82 ± 1 g oil/100 g protein) was seen for PPC–MPC 75:25. These results showed that the blending of plant with milk proteins did not increase the OBC of the samples. The higher OBCs in the SPC and PPC blends (in comparison with the RPC blends) may be an advantage for some functionalities such as for emulsification and applications related to the formulation of breads, cakes and muffins.

3.2.3. Solubility (%)

As shown in Figure 4, the lowest solubility of the plant protein concentrates tested was associated with SPC (43.70 ± 2.33%) and PPC (55.74 ± 2.75%), while the highest was associated with RPC (80.94 ± 0.35%). The solubility varied between the different blends within each plant protein sample highlighting the differences in their interactions with MPC (Figure 4). The mean solubility of the RPC–MPC blends was higher than both the SPC– and PPC–MPC blends. Overall, the RPC–MPC 75:25 blend gave the highest solubility (86.89 ± 1.70%) among the blends ($p < 0.05$). The solubility of PPC was lower than RPC ($p < 0.05$). However, blending PPC with MPC increased its solubility, with the highest being associated with PPC–MPC 50:50 and 25:75. A higher proportion of PPC, however, reduced solubility. SPC had the poorest solubility among the individual plant proteins. Blending SPC–MPC enhanced the overall solubility of SPC. Among the SPC–MPC blends, the highest solubilization was for the SPC–MPC 25:75 blend. The enhanced solubility of some plant protein–MPC blends was evidence for a synergistic relationship between plant proteins and MPC and their interactions with the aqueous phase.

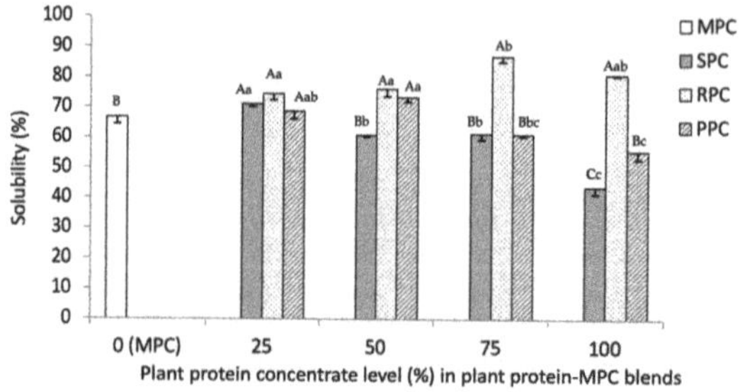

Figure 4. Solubility (%) of soy (SPC), pea (PPC) and rice protein concentrates (RPC) blended with milk protein concentrate (MPC) at different ratios, i.e., 75:25, 50:50 and 25:75. Values represent the mean ± SD ($n = 3$). Different small letters indicate significant differences among different ratios of a constant sample and different capital letters indicate significant differences among different samples at a constant ratio (p value < 0.05).

In addition, the particle size distributions of 5% (*w*/*v*, protein) of the aqueous powder suspensions of each plant protein concentrate and the blends were measured (Section 2.4.4) using laser light scattering (Table 3). The mean Sauter diameter D [3,2 of MPC was 38.55 ± 2.11 μm, which was lower than for SPC and PPC, but was larger than for the RPC sample. The D[3,2] for all tested samples ranged between 8.01 and 101.79 μm, with the RPC–MPC blends having the lowest D[3,2] on average (ranging between 8.01 and 23.33 μm). The low D[3,2] values for the RPC–MPC blends were in line with their higher solubility and lower η_{app} in comparison to PPC– and SPC–MPC blends. However, the polymodal particle size distribution seen in the RPC samples (Figure 5) suggests that these suspensions do not remain stable over time, as those particles may further coalesce. This polydispersity was not observed in either the PPC or the SPC samples. The D[3,2] associated with the SPC–MPC and PPC–MPC blends was in the range of 57.52–101.79 μm. Among the SPC blends, the SPC–MPC 75:25 sample had the highest D[3,2] (88.36 ± 1.93 μm). Similarly, among the PPC blends, the PPC–MPC 75:25 blend had the highest D[3,2] (98.50 ± 3.29 μm). These results showed that the presence of a lower proportion of MPC in SPC–MPC and PPC–MPC blends increased the PS, which is in accordance with the lower solubility observed for these blends.

Table 3. Mean Sauter diameter D[3,2] and specific surface area (SSA) values for soy (SPC), rice (RPC) and pea (PPC) protein concentrates, respectively, blended at 75:25, 50:50 and 25:75 with milk protein concentrate (MPC).

		D[3,2] (μm)	
Plant protein–MPC	SPC–MPC	RPC–MPC	PPC–MPC
0:100	38.55 ± 2.11 ^{a}	38.55 ± 2.11 ^{a}	38.55 ± 2.11 ^{a}
25:75	59.47 ± 1.67 ^{b}	9.42 ± 1.41 ^{d}	58.60 ± 1.91 ^{b}
50:50	59.37 ± 1.85 ^{b}	22.07 ± 1.26 ^{e}	61.35 ± 2.20 ^{b}
75:25	88.36 ± 1.93 ^{c}	12.18 ± 1.44 ^{d}	98.50 ± 3.29 ^{g}
100:0	34.82 ± 1.90 ^{a}	17.04 ± 1.60 ^{f}	47.89 ± 2.03 ^{h}
		SSA (m^2/g)	
0:100	0.16 ± 0.02 ^{AC}	0.16 ± 0.02 ^{AC}	0.16 ± 0.02 ^{AC}
25:75	0.10 ± 0.02 ^{BC}	0.64 ± 0.01 ^{D}	0.10 ± 0.02 ^{BC}
50:50	0.10 ± 0.02 ^{BC}	0.27 ± 0.01 ^{E}	0.10 ± 0.02 ^{BC}
75:25	0.07 ± 0.02 ^{B}	0.49 ± 0.01 ^{F}	0.06 ± 0.03 ^{B}
100:0	0.17 ± 0.02 ^{A}	0.35 ± 0.02 ^{G}	0.13 ± 0.02 ^{C}

Values represent the mean ± SD (n = 3). Different lower-case letters represent significant differences between D[3,2] values and different upper-case letters represent significant differences between SSA ($p < 0.05$) values.

The SSA of the blends was compared (Table 3). The PPC–MPC and SPC–MPC blends gave the lowest SSA. As expected, the RPC–MPC 25:75 blend, which had the lowest D [2,3], presented the highest SSA (0.64 m2/g). In general, particles with smaller sizes and larger surface areas have positive implications for the stability of emulsified foods [29].

According to these results and previous literature, the formation of soluble plant protein–MPC blends depends on the plant protein source, the proportion of the plant protein in the blend and it also depends on the interactions between the plant proteins and CN/WP fractions in MPC.

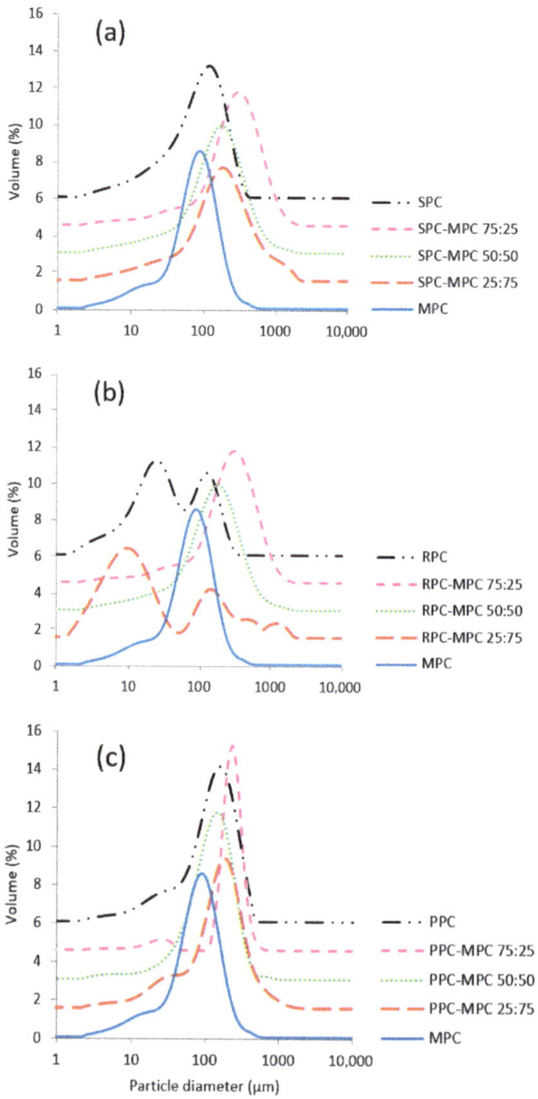

Figure 5. Particle size distribution profiles of (**a**) pea (PPC), (**b**) rice (RPC) and (**c**) soy protein concentrates (SPC) blended at different ratios with milk protein concentrate (MPC) at different ratios, i.e., 75:25, 50:50 and 25:75 (n = 3).

3.2.4. Emulsion Stability (ES%)

Among samples with 100% plant protein, the lowest ES was for the PPC emulsion (51%) and the highest was for SPC (100%). In addition, the ES of MPC was 74%, lower than SPC and RPC and higher than PPC. The emulsification capacity of proteins depends on various factors including their extraction conditions, molecular weight, size, structural rigidity and conformation, and surface charge [30]. The combined effect of these parameters determine the emulsion stability and the impact of one or several parameters may be dominant in certain cases. Based on the results of particle size and solubility measurements (Section 3.2.2), SPC exhibited the smallest particle size after RPC, which results in its faster adsorption to the oil/water interface and therefore, higher emulsification capacity.

In addition a high protein content of this sample, it had the highest apparent viscosity, which could slow down droplet movements in the continuous phase and enhance the emulsion stability. The lowest ES of PPC may be attributed to its considerably larger size compared to other samples as well as its lower protein content compared to other proteins (Table 3). Different mechanisms may also be involved in stabilization of emulsions by different proteins. For instance, investigating the mechanisms of emulsion stabilization of rice protein, pea protein and potato protein in a research study indicated that rice protein forms a physical barrier around oil droplets while pea protein and potato protein from a film layer around the droplets [31].

There were clear differences in the impact of various proportions of plant proteins on ES. Blends which yielded 100% stability after 30 min holding at room temperature were RPC–MPC 50:50, RPC–MPC 75:25, SPC–MPC 75:25, SPC–MPC 100:0 and PPC–MPC 50:50 (Figure 6). The RPC–MPC 25:75 (11%), SPC–MPC 25:75 (21%) and SPC–MPC 50:50 (31%) yielded the lowest ES. Incorporation of 75% plant protein into MPC yielded a high ES in all cases. Incorporation of 50% plant protein to MPC also yielded a high ES in the case of the RPC– and PPC–MPC blends.

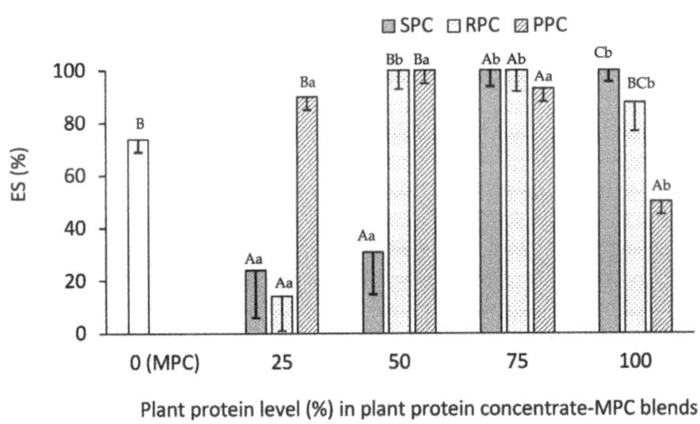

Figure 6. Emulsion stability (ES%) of soy (SPC), rice (RPC) and pea protein concentrates (PPC) blended at different ratios, i.e., 0:100, 25:75, 50:50, 75:25 and 100:0, with milk protein concentrate (MPC) ($n = 3$). Different small letters indicate significant differences among different ratios of a constant sample and different capital letters indicate significant differences among different samples at a constant ratio (p value < 0.05).

The results of ES suggest PPC had the highest extent of interaction with the milk proteins and perhaps the PPC–MPC blends had the highest interfacial energy among the blends, given that the 100% PPC sample exhibited the lowest ES. This considerably improved with the introduction of MPC and was highly stable at a ratio of 50:50 (100%). Furthermore, the PPC–MPC 25:75 emulsion showed higher ES (90%) compared to the other two plant protein samples at the same ratio. This implies a strong interfacial film being generated by the PPC–MPC blend, which was in accordance with a previous report on the emulsion stability of WPI-PPI blends (50:50) [10]. A synergistic action between pea and milk protein has also been highlighted by Hinderink et al., [32] who observed PPI-sodium caseinate and PPI-WP emulsions which remained stable over 14 d, unlike emulsions formed by either PPI and sodium caseinate alone. The rate of adsorption of the blend of dairy proteins (WPI and sodium caseinate) and PPI at the air-water interface was higher than individual proteins showing a synergistic effect arising from blending. In addition, blends of sodium caseinate and PPI had an improved interfacial strength (which is an indication for ES) compared to sodium caseinate alone and thicker films were formed compared to all individual proteins. The emulsion activity of the blends generated with PPI and WPI was recently shown to be higher than the PPI alone [8]. This effect may be associated with reduced flocculation and coalescence by the proteins due to electrostatic interaction between surface protein charges. In addition, Kristensen et al., [8] found that pea proteins are capable of adsorbing to the oil–water interface after introduction to a pre-adsorbed WP interface. On the other hand, Hinderink et al. [33] found that a pre-adsorbed PPC at the oil–water interface can be replaced with β-lactoglobulin. The addition of PPC to infant formula has been reported to have no effect on the emulsion characteristics of the product [34].

It was also found herein that the lower proportion of SPC was less beneficial in the SPC–MPC (25:75) blend system for ES. Increasing the proportion of SPC improved ES. A major increase in ES was observed by increasing the SPC content in the SPC–MPC blend from 50 to 75%. This may be indicative of the high ES of SPC per se, which was previously reported by Molina et al. [35]. Ji et al. [36] reported higher long-term ES in sodium caseinate-SPC emulsions compared to sodium caseinate or SPI stabilized emulsions. Synergic effects on the interfacial strength and viscoelastic film at the air/water interface were also reported for β-conglycinin (7S) and β-lactoglobulin (50:50) compared to the individual proteins [37].

The long-term stability of the prepared emulsions was not evaluated in this work. However, due to the potential application of the emulsions stabilized by plant and animal protein blends in various food products, analysis of long-term stability of emulsification properties under different storage conditions appears to be required.

4. Conclusions

Blending plant proteins with animal proteins is an efficient method to fulfill the increasing global demand for health promoting foods and to decrease the impact of humans on the environment. In addition, the demonstration of new functions for novel protein ingredients compared to existing highly consumed animal-origin proteins is necessary to expand the protein market. Incorporation of plant proteins (SPC, PPC and RPC) with MPC in different ratios in this study showed that these blends may be successfully used for partial replacement of MPC. Multiple blends arising from SPC–, RPC– and PPC–MPC were shown to have functional properties that may be useful in specific food applications. Among the blends, SPC–MPC 25:75, PPC–MPC 50:50 and RPC–MPC 50:50 were shown to be the most suitable in regard to their overall emulsification properties. These findings showed that while various plant protein sources had different emulsion properties, the interactions of these proteins with MPC at certain ratios enhanced their ES. This is advantageous for the partial replacement of MPCs with plant proteins to yield highly stable hybrid emulsions for different applications such as in the manufacture of soups and sauces. Different types of protein–protein interactions take place depending on the characteristics of the individual proteins in the plant protein concentrates and in MPC. Therefore, there is a need for further studies to unravel the nature of these interactions and the potential impact of conventional/novel processing conditions on same. Considering the important impact of solubility on emulsification capability of proteins and the relative low solubility of plant proteins, further investigations should be carried out to introduce the optimal concentration range of plant proteins, resulting in the highest solubility. The long-term stability of the emulsions under various conditions should also be investigated for their application in food industry.

Author Contributions: M.K., conceptualization, formal analysis, methodology, writing—original draft and funding acquisition; S.D., investigation, formal analysis and data curation; J.C., investigation, formal analysis and data curation; C.S., investigation, formal analysis and data curation; S.E., writing—review and editing, data curation and verification; R.J.F., conceptualization, supervision, and editing and funding acquisition. All authors have read and agreed to the published version of the manuscript.

Funding: This research received no external funding.

Institutional Review Board Statement: Not applicable.

Informed Consent Statement: Not applicable.

Data Availability Statement: The data presented in this study are available on request from the corresponding author.

Conflicts of Interest: The authors declare no conflicts of interest.

Abbreviations

The following abbreviations are used in this manuscript:

AA	amino acid
η_{app}	apparent viscosity
CN	casein
DF	diafiltration
ES	emulsion stability
EAA	essential amino acids
MPC	milk protein concentrate
MPI	milk protein isolate
MW	molecular weight
OBC	oil binding capacity
PS	particle size
PP	pea protein
PPC	pea protein concentrate
PPI	pea protein isolate
PAGE	polyacrylamide gel electrophoresis
RP	rice protein
RPC	rice protein concentrate
RPI	rice protein isolate
SDS	sodium dodecyl sulfate
SP	soy protein
SPC	soy protein concentrate
SPI	soy protein isolate
SSA	specific surface area
TS	total solids
UF	ultrafiltration
WHC	water holding capacity
WP	whey protein
WPC	whey protein concentrate
WPH	whey protein hydrolysate
WPI	whey protein isolate

References

1. United Nations. World Population Prospects. 2024. Available online: https://population.un.org/wpp/ (accessed on 1 July 2025).
2. Khalesi, M.; FitzGerald, R.J. Insolubility in milk protein concentrates: Potential causes and strategies to minimize its occurrence. *Crit. Rev. Food Sci. Nutr.* **2022**, *62*, 6973–6989. [CrossRef]
3. Foegeding, E.A.; Davis, J.P. Food protein functionality: A comprehensive approach. *Food Hydrocolloid.* **2011**, *25*, 1853–1864. [CrossRef]
4. Nikbakht Nasrabadi, M.; Sedaghat Doost, A.; Mezzenga, R. Modification approaches of plant-based proteins to improve their techno-functionality and use in food products. *Food Hydrocoll.* **2021**, *118*, 106789–106811. [CrossRef]

5. Khalesi, M.; FitzGerald, R.J. In vitro digestibility and antioxidant activity of plant protein isolate and milk protein concentrate blends. *Catalysts* **2021**, *11*, 787. [CrossRef]
6. Reidy, P.T.; Walker, D.K.; Dickinson, J.M.; Gundermann, D.M.; Drummond, M.J.; Timmerman, K.L.; Fry, C.S.; Borack, M.S. Protein blend ingestion following resistance exercise promotes human muscle protein synthesis. *J. Nutr.* **2013**, *143*, 410–416. [CrossRef]
7. Alves, A.C.; Tavares, G.M. Mixing animal and plant proteins: Is this a way to improve protein techno-functionalities? *Food Hydrocoll.* **2019**, *97*, 105171. [CrossRef]
8. Kristensen, H.T.; Denon, Q.; Tavernier, I.; Gregersen, S.B.; Hammershøj, M.; Van der Meeren, P.; Dewettinck, K.; Dalsgaard, T.K. Improved food functional properties of pea protein isolate in blends and co-precipitates with whey protein isolate. *Food Hydrocoll.* **2021**, *113*, 106556. [CrossRef]
9. Le Roux, L.; Mejean, S.; Chacon, R.; Lopez, C.; Dupont, D.; Deglaire, A.; Nau, F.; Jeantet, R. Plant proteins partially replacing dairy proteins greatly influence infant formula functionalities. *LWT Food Sci. Technol.* **2020**, *120*, 108891. [CrossRef]
10. Ho, K.K.H.Y.; Schroen, K.; Martín-Gonzalez, M.F.S.; Berton-Carabin, C.C. Synergistic and antagonistic effects of plant and dairy protein blends on the physicochemical stability of lycopene-loaded emulsions. *Food Hydrocoll.* **2018**, *81*, 180–190. [CrossRef]
11. Beliciu, C.M.; Moraru, C.I. Physico-chemical changes in heat treated micellar casein–Soy protein mixtures. *LWT Food Sci. Technol.* **2013**, *54*, 469–476. [CrossRef]
12. Mao, C.; Xin, M.; Wu, W.; Wang, S.; Wu, J.; Cheng, Y.; Ma, H. Interfacial characteristics and digestive kinetics of emulsions fabricated by composite potato and whey protein. *Int. J. Biol. Macromol.* **2025**, *308*, 142446. [CrossRef]
13. Wu, M.; He, X.; Feng, D.; Li, H.; Han, D.; Li, Q.; Zhao, B.; Li, N.; Liu, T.; Wang, J. The emulsifying properties, In vitro digestion characteristics and storage stability of high-pressure-homogenization-modified dual-protein-based emulsions. *Foods* **2023**, *12*, 4141. [CrossRef]
14. Zhang, X.; Zhang, S.; Xie, F.; Han, L.; Li, L.; Jiang, L.; Qi, B.; Li, Y. Soy/whey protein isolates: Interfacial properties and effects on the stability of oil-in-water emulsions. *J. Sci. Food Agric.* **2021**, *101*, 262–271. [CrossRef]
15. Hinderink, E.B. Food Emulsions Stabilised by Blends of Plant and Dairy Proteins. Ph.D. Thesis, Wageningen University and Research, Wageningen, The Netherlands, 2021.
16. Hu, T.; Chen, J.; He, X.; Tang, Y.; Sun, J.; Liu, C.; Dai, T. Complex plant protein prepared from rice protein and pea protein: Improve the thermal stability of betanin. *Food Res. Int.* **2023**, *164*, 112341. [CrossRef]
17. Zhao, H.; Shen, C.; Wu, Z.; Zhang, Z.; Xu, C. Comparison of wheat, soybean, rice, and pea protein properties for effective applications in food products. *J. Food Biochem.* **2020**, *44*, 13157. [CrossRef] [PubMed]
18. Khalesi, M.; FitzGerald, R.J. Physicochemical properties and water interactions of milk protein concentrate with two different levels of undenatured whey protein. *Colloids Surf. A Physicochem. Eng. Asp.* **2021**, *629*, 127516. [CrossRef]
19. Kaspchak, E.; Silveira, J.L.; Igarashi-Mafra, L.; Mafra, M.R. Effect of antinutrients on heat-set gelation of soy, pea, and rice protein isolates. *J. Food Sci. Technol.* **2020**, *57*, 4201–4210. [CrossRef] [PubMed]
20. Webb, M.F.; Naeem, H.A.; Schmidt, K.A. Food protein functionality in a liquid system: A comparison of deamidated wheat protein with dairy and soy proteins. *J. Food Sci.* **2002**, *67*, 2896–2902. [CrossRef]
21. Nadathur, S.R.; Wanasundara, J.P.D.; Scanlin, L. *Sustainable Protein Sources*; Elsevier: Amsterdam, The Netherlands, 2017.
22. Tarrega, A.; Ramírez-Sucre, M.O.; Vélez-Ruiz, J.F.; Costell, E. Effect of whey and pea protein blends on the rheological and sensory properties of protein-based systems flavoured with cocoa. *J. Food Eng.* **2012**, *109*, 467–474. [CrossRef]
23. Yerramilli, M.; Longmore, N.; Ghosh, S. Improved stabilization of nanoemulsions by partial replacement of sodium caseinate with pea protein isolate. *Food Hydrocolloid.* **2017**, *64*, 99–111. [CrossRef]
24. Dapčević-Hadnađev, T.; Dizdar, M.; Pojić, M.; Krstonošić, V.; Zychowsk, L.M.; Hadnađev, M. Emulsifying properties of hemp proteins: Effect of isolation technique. *Food Hydrocoll.* **2019**, *89*, 912–920. [CrossRef]
25. Onwulata, C.I.; Tunick, M.H.; Mukhopadhyay, S. Flow behavior of mixed-protein incipient gels. *Int. J. Food Prop.* **2014**, *17*, 1283–1302. [CrossRef]
26. Chew, J.H.; Liu, W.; Fu, N.; Gengenbach, T.; Chen, X.D.; Selomulya, C. Exploring the drying behaviour and particle formation of high solids milk protein concentrate. *J. Food Eng.* **2014**, *143*, 186–194. [CrossRef]
27. Chavana, U.D.; McKenzie, D.B.; Shahidi, F. Functional properties of protein isolates from beach pea (*Lathyrus maritimus* L.). *Food Chem.* **2001**, *74*, 177–187. [CrossRef]
28. Reinkensmeier, A.; Bußler, S.; Schlüter, O.; Rohn, S.; Rawel, H.M. Characterization of individual proteins in pea protein isolates and air classified samples. *Food Res. Int.* **2015**, *76*, 160–167. [CrossRef]
29. Malaki Nik, A.; Tosh, S.M.; Woodrow, L.; Poysa, V.; Corredig, M. Effect of soy protein subunit composition and processing conditions on stability and particle size distribution of soymilk. *LWT Food Sci. Technol.* **2009**, *42*, 1245–1252. [CrossRef]
30. Sagis, L.M.; Yang, J. Protein-stabilized interfaces in multiphase food: Comparing structure-function relations of plant-based and animal-based proteins. *Curr. Opin. Food Sci.* **2022**, *43*, 53–60. [CrossRef]
31. da Silva, A.M.M.; Almeida, F.S.; Sato, A.C.K. Functional characterization of commercial plant proteins and their application on stabilization of emulsions. *J. Food Eng.* **2021**, *292*, 110277. [CrossRef]

32. Hinderink, E.B.A.; Sagis, L.; Schroën, K.; Berton-Carabin, C.C. Behavior of plant-dairy protein blends at air-water and oil-water interfaces. *Colloids Surf. B.* **2020**, *192*, 111015. [CrossRef]
33. Hinderink, E.B.A.; Sagis, L.; Schroën, K.; Berton-Carabin, C.C. Sequential adsorption and interfacial displacement in emulsions stabilized with plant-dairy protein blends. *J. Colloid Interface Sci.* **2021**, *583*, 704–713. [CrossRef]
34. Ju, Z.Y.; Hettiarachchy, N.S.; Rath, N. Extraction, denaturation and hydrophobic properties of rice flour proteins. *J. Food Sci.* **2006**, *66*, 229–232. [CrossRef]
35. Molina, E.; Papadopoulou, A.; Ledward, D.A. Emulsifying properties of high pressure treated soy protein isolate and 7S and 11S globulins. *Food Hydrocoll.* **2001**, *15*, 263–269. [CrossRef]
36. Ji, J.; Zhang, J.; Chen, J.; Wang, Y.; Dong, N.; Hu, C.; Chen, H.; Li, G.; Pan, X.; Wu, C. Preparation and stabilization of emulsions stabilized by mixed sodium caseinate and soy protein isolate. *Food Hydrocoll.* **2015**, *51*, 156–165. [CrossRef]
37. Pizones Ruiz-Henestrosa, V.M.; Martinez, M.J.; Carrera Sánchez, C.; Rodríguez Patino, J.M.; Pilosof, A.M.R. Mixed soy globulins and β-lactoglobulin systems behaviour in aqueous solutions and at the air–water interface. *Food Hydrocoll.* **2014**, *35* (Suppl. C), 106–114. [CrossRef]

foods

MDPI

Article

Functional, Biological and Nutritional Properties of Protein Fraction Isolated from *Yarrowia lipolytica* Biomass

Marek Szołtysik, Anna Mandecka *, Marcelina Maciejewska, Anna Dąbrowska and Marek Nowak

Department of Development Functional Food Products, Faculty of Biotechnology and Food Science, Wroclaw University of Environmental and Life Sciences, 50-366 Wroclaw, Poland; marek.szoltysik@upwr.edu.pl (M.S.); marcelina.maciejewska@upwr.edu.pl (M.M.); anna.dabrowska@upwr.edu.pl (A.D.); marek.nowak1@upwr.edu.pl (M.N.)
* Correspondence: anna.mandecka@upwr.edu.pl; Tel.: +48-667-421-235

Abstract

This study evaluated the nutritional, functional, biological, and sensory potential of proteins derived from *Yarrowia lipolytica* biomass and their enzymatic hydrolysates for food applications. Three strains were cultivated under bioreactor conditions, with strain JII1c selected for its superior biomass yield and protein content. Its amino acid composition was rich in lysine and branched-chain amino acids, with protein quality indices (CS = 37.8%, EAAI = 36.17%) confirming value in plant-based diets. Proteins were isolated and hydrolysed using a non-commercial serine protease from *Cucurbita ficifolia*, which enhanced solubility (NSI: 19.4 → 49.2%), water and oil absorption, and emulsion stability. Hydrolysates showed notable biological activities, including ACE (71.8%), DPP-IV (52.3%), and α-glucosidase (67.4%) inhibition, indicating potential metabolic benefits. Sensory evaluation of extrudates confirmed improvements in aroma, texture, and flavour when hydrolysates were incorporated. The use of a plant-derived protease demonstrates a sustainable approach to producing bioactive peptides. *Y. lipolytica* hydrolysates emerge as promising clean-label ingredients that combine nutritional quality with techno-functional performance, supporting their integration into health-oriented and sustainable food products.

Keywords: *Yarrowia lipolytica* yeast; enzyme hydrolysis; bioactive peptides; functional properties

1. Introduction

The global demand for high-quality protein is rising due to protein–energy malnutrition, amino acid imbalance, climate change, and shifting dietary preferences. Conventional protein sources such as meat, dairy, and fish, though rich in essential amino acids, contribute to environmental degradation through greenhouse gas emissions, land use, and water consumption [1,2]. Despite overall growth in food output, many populations still lack access to adequate diets, reflecting deficiencies of food systems that prioritize yield and cost-efficiency over nutritional quality [3]. Meanwhile, vegetarian, vegan and flexitarian trends are driving demand for novel protein ingredients that are both sustainable and nutritionally balanced. In the United States, plant-based foods reached 61% household penetration in 2021, with 19% buying meat substitutes [2,4–7].

Alternative proteins from plants, insects, microalgae and fungi are gaining attention for their environmental and functional potential [8,9]. Among microbial sources, *Yarrowia lipolytica* is particularly promising, offering up to 60% protein per dry weight when cultivated under optimised bioreactor conditions using agro-industrial by-products [10]. Its proteins exhibit a complete amino acid profile, with high levels of lysine and leucine, surpassing even egg protein in lysine content [5,11]. *Y. lipolytica* has been recognized as safe for food and feed applications, holding GRAS status in the United States and approval as a novel food by the European Food Safety Authority (EFSA, 2019), confirming its safety, metabolic versatility, and high productivity [12]. Additionally, its cell wall provides bioactive polysaccharides and B vitamins such as riboflavin and B12, aligning with clean-label and sustainability trends [5,13,14].

This yeast also supports circular economy strategies by valorising waste streams, including bread, brewer's spent grain and lignocellulosic residues, without extensive pretreatment [15–17]. Such processes reduce environmental burden while producing protein-rich biomass for food and feed applications.

Protein hydrolysis is an effective approach to enhance the biofunctionality of yeast proteins, releasing bioactive peptides (BAPs) with antioxidant, antihypertensive, antimicrobial and anti-inflammatory effects [18]. Functional properties of hydrolysates depend on substrate composition and processing conditions [19]. Plant-derived enzymes are particularly attractive for sustainable applications. A serine protease from *Cucurbita ficifolia* (Asian pumpkin) has been shown to generate peptides with strong antioxidant and functional properties and has been successfully applied to substrates including casein and insect protein [20–23]. Recent studies confirm that *Y. lipolytica* hydrolysates obtained via enzymatic treatment display potent antioxidant activity and functionality as flavour enhancers, emulsifiers and foaming agents. For example, Gottardi et al. (2022) demonstrated that hydrolysates obtained with *Y. lipolytica* strain YL2 showed strong DPPH radical scavenging activity, reaching 86.4% after 72 h of incubation, confirming the high bioactive potential of peptides derived from this yeast [5,14,24,25].

The aim of this study was to evaluate the potential of *Y. lipolytica* proteins and their enzymatic hydrolysates in food applications. The research focused on optimising cultivation and biomass processing for high protein yield, followed by enzymatic hydrolysis to generate bioactive peptide fractions. The nutritional, functional and sensory properties of both protein isolates and hydrolysates were assessed to explore their value in plant-based formulations.

2. Materials and Methods

2.1. Substrates

The material used for the study was the biomass of *Yarrowia lipolytica* yeast, which was obtained from bioreactor cultures (*Yarrowia lipolytica* yeast was obtained from the cultures collection of the Department of Biotechnology and Food Microbiology, Wroclaw University of Environmental and Life Sciences). The protein was then isolated from this biomass. Three distinct yeast strains were utilised for the culture of *Y. lipolitica*: JII1c, JII1a, and PII6a. The medium used for the reactor cultures was of plant origin, i.e., it was a vegan product.

2.2. Bioreactor Culture

Cultures were grown in a BioFlow 310 bioreactor (New Brunswick Scientific GmbH, Nurtingen, Germany) with a total volume of 7 L and a working volume of 4 L at a temperature of 28 °C. Aeration was maintained at 1 vvm, and agitation was set at 500 rpm for 48–72 h. The culture medium consisted of (g/L): yeast extract (1.7), casein (4.0), refined fatty acids (10.0), KH_2PO_4 (0.5), $MgSO_4{\cdot}7H_2O$ (0.25) and NH_4Cl (1.0). Inoculum propagation was conducted sequentially in three bioreactors of increasing volume. In the first stage, a 7 L BioFlow 310 bioreactor (working volume: 4 L) was operated under the same temperature, aeration (1 vvm), and agitation (500 rpm) conditions for 48 h. The second stage involved a 50 L MPPF bioreactor (New Brunswick Scientific Co., Inc., Edison, NJ, USA) with a working volume of 4 L, maintained at identical operating conditions. In the third stage, cultivation was continued in the same 50 L bioreactor with an increased working volume, while aeration was reduced to 0.6 vvm and agitation adjusted to 350 rpm. The resulting concentrate was subjected to spray-drying with the addition of 3% (*w*/*w*) methylcellulose as a carrier. Spray-drying was performed at an inlet air temperature of 140 °C and an outlet air temperature of 88 °C, with a feed flow rate of 2.8 L/h and an air pressure of 1.1 bar. Biomass was harvested based on optical density. Optical density was reported based on the standard curve, for which extinction was measured for a specific amount expressed as [log10 CFU/mL]. Three replicates were performed for each strain.

2.3. Nutritional Quality Assessment of Biomasses

The obtained biomasses were analysed for their content of dry matter, protein, carbohydrates (including dietary fibre and sugars), fat and fatty acids, and both macro- and microelements [26–32]. B vitamins were determined by HPLC after enzymatic digestion and saponification, using a C18 column with UV/fluorescence detection [33–38]. Determination of the content of sodium was carried out in an acetylene/air flame by atomic emission spectrometry using the SpectraAA atomic absorption spectrometer with the flame attachment AA240FS (Varian). The methods were validated using certified reference material; the measurement uncertainty was estimated at 5%. Mineralization was performed in accordance with the Polish Standard PN-EN 13805:2003 "Foodstuffs. Determination of trace elements. Pressure mineralization" [39].

The amino acid content was analysed by reversed-phase high-performance liquid chromatography (RP-HPLC). Samples (500 µL) were solubilised in Tris/HCl buffer with urea and 2-mercaptoethanol, loaded on a Zorbax XDB-C18 column (250 × 4.5 mm, Agilent Technologies, Inc., Santa Clara, CA, USA), and separated at 1 mL/min, 30 °C, with gradient elution (0–100% acetonitrile/TFA); detection at 230 nm [40].

To assess the nutritional quality of the yeast biomasses, the method described by Liang et al. was applied [41]. Essential amino acid (EAA) profiles were analysed and compared to a reference protein (whole egg). Two parameters were calculated: the Chemical Score (CS) and the Essential Amino Acid Index (EAAI).

The CS was determined as the ratio of each essential amino acid present in the protein sample to its corresponding content in whole egg protein, multiplied by 100. The final CS was expressed as the geometric mean of these individual ratios, according to the equation:

$$\text{Chemical score} = \left(\frac{\text{EAA in sample}}{\text{EAA in whole egg}}\right)\text{x}100 \tag{1}$$

The essential amino acid index (EAAI) reflects the overall balance of essential amino acids relative to the reference protein and provides a holistic metric for protein quality. The reference amino acid pattern of whole egg protein was adopted from FAO/WHO/UNU (1985) [42]. The EAAI was calculated as the geometric mean of the ratios of each essential amino acid in the sample to its corresponding content in whole egg protein. The calculation followed the formula:

$$\text{EAAI} = 100\text{x}\sqrt[n]{\frac{\text{Lys}_\text{p}}{\text{Lys}_\text{s}}\text{x}\frac{\text{Trp}_\text{p}}{\text{Trp}_\text{s}}\text{x}\ldots\text{x}\frac{\text{Val}_\text{p}}{\text{Val}_\text{s}}} \tag{2}$$

where

p—refers to the protein sample;

s—refers to the standard (whole egg protein);

n—is the number of essential amino acids considered (with Met + Cys and Phe + Tyr treated as combined pairs).

The energy value of biomass was calculated using the Atwater general factors (4 kcal/g for protein, 9 kcal/g for fat, and 4 kcal/g for carbohydrates), based on the measured contents of macronutrients, according to FAO/WHO recommendations and Codex Alimentarius guidelines [43].

2.4. Microbiological Analysis

The microbiological safety of the dried *Yarrowia lipolytica* biomass was evaluated according to international standards. Total aerobic mesophilic bacteria (TAMC) were determined using ISO 4833-1:2013 [44]. Total yeast and mould count (TYMC) was measured according to ISO 21527-2:2008 [45]. The presence of viable *Yarrowia lipolytica* cells was confirmed using YPD agar (yeast extract–peptone–dextrose) incubated at 28 °C for 48 h, with colony morphology verified microscopically and through Gram staining. Coliform bacteria were detected based on ISO 4832:2006 [46]. *Salmonella* spp. detection followed ISO 6579-1:2017 [47]. Results were expressed as presence or absence in 25 g of sample.

2.5. Protein Extraction from JII1c Biomass

The protein isolates (PI) were obtained from *Yarrowia lipolytica* biomass JII1c using a two-step procedure. Initially, the yeast biomass, suspended in water, was subjected to acid hydrolysis to disrupt the cell walls and release intracellular components. In the first step, NaOH was used at a concentration of 1 M (t = 60 min), and in the second step, the 1 M HCl was used (t = 15 min). Following cell disruption, proteins were extracted under alkaline conditions (pH = 12) to increase their solubility. Subsequently, protein precipitation was induced by adjusting the pH to the isoelectric point (pH = 4.5), as described by Shetty K. J. & Kinsella (1979) [48]. During both steps, mechanical stirring was used. The protein content was determined post-precipitation by the Kjeldahl method. The efficiency of the entire process was estimated at approximately 55–60%.

2.6. Enzymatic Hydrolysis (EH)

2.6.1. Enzyme Applied in the Study

Serine proteinase was isolated from Asian pumpkin (*Curcubita ficifolia*) by the method according to Dryjanski et al., 1990 [49] Proteolytic activity was determined using 2% casein in Tris/HCl (pH 8.6), incubated for 10 min at 35.5 °C. The reaction was stopped with 5% TCA, centrifuged at 5500× *g* (RCF) for 10 min, and absorbance was measured at λ = 280 nm. One unit was defined as a 0.1 increase in absorbance.

2.6.2. Hydrolysis Procedure

Protein isolates (PI) from *Yarrowia lipolytica* JII1c biomass were subjected to enzymatic hydrolysis (EH) using pepsin 107,192 (Sigma-Aldrich, Poznań, Poland) and and non-commercial serine protease extracted from *Cucurbita ficifolia*. Enzymes were added at 150 enzyme units (U) per mg of protein to 1% substrate protein solution.

Hydrolysis was conducted at 37 °C, and samples as protein hydrolysates (PHs) were collected at 0, 5, 12, and 24 h for further analyses. The enzyme was thermally inactivated at 100 °C. After inactivation, the hydrolysates were centrifuged (5000× *g*; 20 min) and then frozen. The hydrolysates were captured at −18 °C and resuspended at room temperature before analysis.

The control consisted of protein isolates (PIs) not treated with enzymes but subjected to the same thermal conditions and incubation times as the enzymatic samples. The hydrolysis conditions were established based on previously published studies describing the catalytic characteristics *of Cucurbita ficifolia* serine protease [20–23].

2.6.3. Degree of Hydrolysis (DH) [%]

The degree of hydrolysis (DH) was determined according to the method of Silvestre (1996), based on the concentration of peptides soluble in 5% trichloroacetic acid (TCA) [50]. Briefly, 1 mL of hydrolysate was mixed with 1 mL of 10% TCA and allowed to stand for 1 h to precipitate unhydrolyzed proteins. The samples were then centrifuged at 5000× *g* for 15 min at 5 °C. Protein concentration in the obtained supernatant was measured spectrophotometrically at wavelengths of λ = 280 and λ = 235 nm. The degree of hydrolysis was expressed as the relative increase in soluble peptide content after enzymatic treatment compared to the control sample.

2.7. *Determination of Antioxidant Activity*

2.7.1. DPPH

Antioxidant activity was determined by the modified Yen & Chen (1995) method by adding 1 mL of 96% ethanol and 500 μL of a 0.3 mM ethanolic DPPH (2,2-di(4-tertoctylphenyl)-1-picrylhydrazyl) radical solution to 1 mL of peptide solution in Tris-HCl buffer at pH 7.0, with mixing [51]. Absorbance was measured at λ = 517 nm after 30 min of incubation. The antioxidant activity of a 1 mg/mL peptide solution was determined on the basis of a standard curve prepared for Trolox and expressed as [μM Trolox/mg].

2.7.2. Ferric Reducing Antioxidant Power Assay (FRAP)

Antioxidant activity was determined as the ability of the hydrolysate to reduce the Fe(III) to Fe(II) ions in reaction with TPTZ (2,4,6-Tris(2-pyridyl)-s-triazine) [52]. To 1 mL of an aqueous peptide solution was added 3 mL of a working solution containing 1A:1B:10C, where A (10 mM TPTZ in 40 mM HCl), B (20 mM $FeCl_3 \times 6 H_2O$), C (0.3 M acetate buffer pH 3.6 (3.1 g $C_2H_3NaO_2 \times 3H_2O$, 16 mL acetic acid/1 L H_2O). The samples were incubated for 10 min at room temperature, and then the absorption was measured at λ = 593 nm. Antioxidant activity was expressed as the ability to reduce the oxidation state of iron ions—Fe(III) to Fe(II). The concentration of Fe(II) ions in the sample was calculated based on a standard curve prepared for specific concentrations of $FeSO_4$ solution. The reducing capacity of enzymatic hydrolysates and peptides was calculated per 1 mg of protein. The results were expressed as [g Fe^{3+}/mg].

2.7.3. Fe(II) Ion Chelation

Chelation of iron ions was determined by colorimetric measurement of Fe(II) not bound by hydrolysate in a reaction mixture using ferrozine (3-(2-pyridyl)-5,6-diphenyl-1,2,4-triazine-p,p′-disulfonic acid monosodium salt hydrate) [53]. To 250 μL of appropriately diluted peptide solution was added 1250 μL of H_2O and 110 μL of 1 mM $FeCl_2$ (9.94 mg $FeCl_2$ in 50 mL H_2O). After 2 min of incubation, 1 mL of 0.5 mM ferrozine solution (12.31 mg ferrozine in 50 mL H_2O) was added. The samples were kept for 10 min at room temperature, after which absorbance was measured at λ = 562 nm. Simultaneously, a blank sample was performed, in which 1000 μL of distilled H_2O and a reagent sample (110 μL of 1 mM $FeCl_2$, 1190 μL H_2O, 1000 μL H_2O) were added. Iron ion chelating activity was expressed as micrograms of bound $FeCl_2$ per milligram of protein [μg Fe^{2+}/mg], based on a standard curve prepared with $FeCl_2$ as standard.

2.7.4. ACE Inhibitory Activity

The ACE inhibitory activity was determined according to the method of Hernández-Ledesma et al., with modifications made for the purpose of the study [54]. Briefly, 20 μL of each sample was added to 0.1 mL of 0.1 M potassium phosphate buffer (pH 8.3) containing 0.3 M NaCl and 5 mM Hip-His-Leu (HHL) as the substrate. After the addition of 5 milliunits (mU) of ACE (Angiotensin-Converting Enzyme, Sigma Aldrich, >2.0 U/mg protein), the reaction mixture was incubated at 37 °C for 30 min. The reaction was stopped by adding 0.1 mL of 1 M HCl. The hippuric acid formed was extracted with ethyl acetate, evaporated by heating at 95 °C for 10 min, redissolved in distilled water, and measured spectrophotometrically at $\lambda = 228$ nm. All assays were performed in triplicate. The inhibitory activity was expressed as the percentage of ACE inhibition at a given peptide nitrogen concentration, and the IC_{50} value (the concentration of peptides required to inhibit 50% of ACE activity) was calculated.

2.7.5. α-Glucosidase Inhibitory Activity

According to Yu et al., peptides were incubated with α-glucosidase and substrate (pNPG) [55]. To 610 μL of 0.1 M potassium phosphate buffer, pH 6.8, were added 10 μL of 3 mM reduced glutathione solution, 5 μL of α-glucosidase (10 U/mL), and 10 μL of peptide solution. The mixture was preincubated for 20 min at 37 °C, after which the reaction was initiated by adding 10 μL of the substrate p-NPG (p-nitrophenyl β-D-glucopyranoside) (10 mM). The reaction was continued for 30 min at 37 °C. After this time, 650 μL of 1 M Na_2CO_3 was added to terminate the process. The amount of released product, p-nitrophenol, was determined spectrophotometrically at $\lambda = 410$ nm. The enzyme inhibitory activity (IC_{50}) was expressed as the amount of substance required to half-inhibit (IC_{50}) the activity of α-glucosidase under the conditions described above.

2.7.6. DPP-IV Inhibitory Activity

Activity was determined according to Giovanni Tulipano's method [56]. The test material was suspended in 0.1 M Tris-HCl buffer, pH 8.0. The sample (25 μL) was preincubated with the same volume of Gly-Pro-p-nitroanilide substrate (1.6 mM) at 37 °C for 10 min. Then, 50 μL of DPP-IV (0.01 U/mL, in 0.1 M Tris-HCl buffer, pH 8.0) was added to the mixture and incubated at 37 °C for 60 min. The reaction was stopped by adding 100 μL of 1 M acetate buffer, pH 4.0. The released p-nitroanilide hydrolysis product was measured at $\lambda = 405$ nm. The DPP-IV enzyme inhibitory activity (IC_{50}) was expressed as the amount of substance required to half-inhibit the DPP-IV enzyme activity under the conditions described above.

2.8. Functional Properties Measurement

2.8.1. Water Absorption Capacity (WAC)

The method was based on Timilsena et al. (2016) [57]. Approximately 1 g of the protein preparation was weighed into a test tube and mixed with 20 mL of distilled water. The mixture was shaken using a laboratory shaker. After 15 min, it was shaken again for 60 s and then centrifuged at 4500× *g* (RCF) for 15 min (Rotofix 32A; Merazet, Poland). Unabsorbed water was carefully decanted, and remaining droplets were removed with blotting paper. The solid phase was dried at 50 °C for 30 min. Water binding capacity (W [g H_2O/g]) was calculated as:

$$W = \frac{C - B}{A} \tag{3}$$

where

C—weight of tube with dried sediment (g);
B—weight of empty tube (g);
A—sample weight (g).

2.8.2. Oil Absorption Capacity (OAC)

The procedure followed that of Wu et al. (2009) [58]. One g of protein preparation was mixed with 15 mL of rapeseed oil and shaken. After 30 min, the mixture was at 4000× *g* (RCF) for 10 min. The oil binding capacity (X) was calculated as:

$$X = \frac{\mathrm{V} - \mathrm{L}}{\mathrm{a}} \tag{4}$$

where

V—volume of oil used (mL);

L—volume of unabsorbed oil (mL);

a—weight of the sample, calculated per 100 g of dry matter.

2.8.3. Nitrogen Solubility Index (NSI)

According to Achouri et al. (2012) [59], 200 mg of protein preparation was mixed with 15 mL of distilled water. The pH was adjusted to 2, 4, 6, 8, 10, or 12 using 0.5 M NaOH or 0.5 M HCl. Samples were shaken at room temperature for 30 min, then centrifuged at 4500× *g* (RCF, ≈10,000 rpm) for 15 min (MPW-351 R centrifuge, MPW, Warsaw, Poland, 10,000 rpm). From the supernatant, 10 g was taken for total nitrogen determination using the Kjeldahl method. The total protein content was calculated using a nitrogen-to-protein conversion factor of 6.25, in triplicate. Protein solubility (PS) was calculated as:

$$\mathrm{PS} = \frac{\mathrm{A}}{\mathrm{B}}\ \mathrm{x}\ 100\% \tag{5}$$

where

A—protein content in supernatant (g);

B—protein content in 200 mg of preparation.

2.8.4. Emulsion Stability

Hydrolysates at 1%, 2%, and 3% concentrations were mixed with 10 mL of water and homogenized until uniform. Pre-emulsification was performed using a lab homogenizer for 2 min at 16,000 rpm (T10 basic ULTRA-TURRAX, IKA WerkeGmbH&Co., Baden-Wurttemberg, Germany). Subsequently, the homogenizer was operated for 20 s, during which a quantity of 10 mL of soybean oil was added. Homogenization continued until 2 min had passed from the first drop of oil (total time: 180 s). The emulsion was transferred to a 10 mL polypropylene tube (14 mm internal diameter). Phase separation (sedimentation) was observed at 30 min, 1 h, 2 h, and 3 h. Emulsion stability (SE) was calculated as:

$$\mathrm{SE} = \frac{\mathrm{a}}{\mathrm{b}}\ \mathrm{x}\ 100 \tag{6}$$

where

a—height of emulsified layer [cm];

b—total emulsion height [cm].

2.8.5. Foam Stability

Into a 1 L beaker, 100 mL of fresh egg white was mixed with hydrolysate (1%, 2%, 3%). The mixture was whipped using a kitchen mixer at 800 rpm for 10 min. The liquid at the bottom was poured into a graduated cylinder. The beaker with foam was left at ~20 °C for 30 min. After 30 min, drainage liquid from the foam was collected and combined with the initial bottom phase for total drainage volume measurement. Foam stability was expressed in millilitres as the total drainage volume and as a percentage [%] of retained foam volume. The control sample contained only fresh egg white, without the addition of any test substances.

2.9. Protein Extrudates

2.9.1. Production

The protein isolates and their hydrolysates obtained from *Yarrowia lipolytica* JII1c biomass were processed using a laboratory twin-screw extruder (AEV 650, Brabender). The extrusion was conducted according to the methodology of Rytel, under controlled conditions, with a screw compression ratio of 4:1, a screw speed of 180 revolutions per minute, and a load range of 4.5 to 7 amperes [60]. A die with a diameter of 4 mm was used for shaping the extrudates. To assess the impact of processing temperature, the extrusion was performed at three different temperatures: 160 °C, 170 °C, and 180 °C. These parameters were selected to simulate industrial processing conditions for the development of high-protein, plant-based meat analogues.

2.9.2. Sensory Analysis

The sensory evaluation was conducted by a trained panel (n = 8) in accordance with ISO 13299:2016 guidelines [61]. The assessed attributes included taste, odour, texture, saltiness, and appearance. Evaluations were performed under controlled laboratory conditions (20 ± 1 °C, neutral lighting, individual booths). Samples of extrudates were coded with random three-digit numbers and presented in randomized order to avoid positional bias. Panelists rinsed their mouths with water between evaluations. Each attribute was rated on a 5-point hedonic scale, where 1 indicated "very poor" and 5 indicated "excellent." Results were expressed as mean values. Bioethics approval for the study was granted by the Rector's Committee for Ethics in Scientific Research, Wroclaw University of Environmental and Life Sciences (Resolution No. N0N00000.020.1.8.4.2024, dated 21 October 2024).

2.10. Statistical Analysis

The analysis was performed with the use of the "STATISTICA 13 PL" program by StatSoft Inc., Tulsa, OK, USA. All determinations were performed in triplicate and presented as average value (X) with standard deviation (SD). Assessment of the significant differences between average values in analysed groups was conducted by variance analysis (ANOVA), followed by a Duncan multiple range test. The level of statistical significance was set at $p < 0.05$.

3. Results

A significant productivity difference was observed between the three *Yarrowia lipolytica* strains cultivated under bioreactor conditions (Figure 1). The highest determined cell count was observed in strain JII1c, which reached 9.62 $\log_{10}$ CFU/mL, indicating its strong potential for high-yield biomass production. Conversely, strain JII1a exhibited the lowest productivity, measuring 7.86 $\log_{10}$ CFU/mL, while PII6b demonstrated an intermediate value of 8.04 $\log_{10}$ CFU/mL (1II1a vs. JII1c, $p = 0.00011$; 1II1a vs. PII6b, $p = 0.04193$; JII1c vs. PII6b, $p = 0.00023$). The observed differences in cell density were supported by standard deviation bars, suggesting biological variation across replicates. The markedly higher cell count of JII1c reinforces its selection for subsequent protein isolation, as it combines both high biomass accumulation and favourable compositional traits.

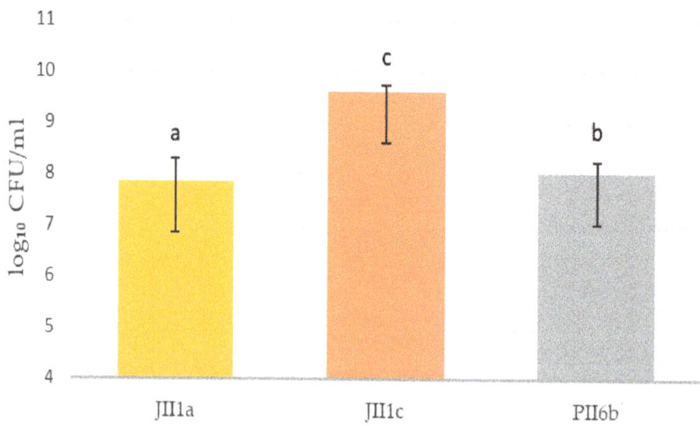

Figure 1. Biomass productivity of selected yeast strains of *Y. lipolytica* in bioreactor cultures [$\log_{10}$ CFU/mL]. Small letters (a–c) indicate statistically significant values: a vs. b vs. c ($p < 0.05$), CFU—colony forming units.

3.1. Biomass Yield and Nutritional Composition

The dried biomass of *Yarrowia lipolytica* JII1c appeared as a light beige, free-flowing powder with a characteristic yeast-like odour. Microbiological analysis confirmed its safety, showing <10 CFU/mL viable cells, $\leq 5 \times 10^3$ CFU/mL total aerobic microorganisms, $\leq 10^2$ CFU/mL yeasts and moulds, and the absence of coliform bacteria and *Salmonella* spp. in 25 g of the sample.

The Nutritional characteristic of dried biomass from *Yarrowia lipolytica* JII1c strain was presented in Table 1. The chemical composition revealed a high dry matter content (95.11%) and a substantial protein level (43.12%), thus confirming the strain's potential as a protein-rich ingredient. The carbohydrate content constituted 32.34% of the total composition, with the predominant proportion being dietary fibre (32.32%), and negligible levels of sugars (<0.20%), thereby substantiating its compatibility for low glycaemic formulations. The fat content was moderate (7.03%) and primarily composed of monounsaturated (4.05%) and polyunsaturated (3.30%) fatty acids, with saturated fats present at minimal levels (0.50%). The ash content (11.00%) indicated an elevated mineral load, while the low moisture content (4.89%) ensured the desired shelf stability. The salt content was found to be 4.62 g/100 g.

The biomass also proved to be a source of micronutrients relevant to human nutrition. It contained a complex of B-group vitamins, including riboflavin, biotin, folic acid, and vitamin B12—compounds of particular importance in plant-based diets. The mineral profile included considerable amounts of phosphorus, potassium, and sodium, alongside nutritionally relevant levels of iron, zinc, magnesium, and other trace elements.

The amino acid profile indicated the presence of all essential and non-essential amino acids, with lysine (31.1 g/kg), leucine (31.2 g/kg), and valine (25.1 g/kg). The content of methionine + cysteine (9.3 g/kg), tryptophan (5.1 g/kg), and histidine (8.7 g/kg) further emphasized the high nutritional quality of the protein. Furthermore, the presence of gamma-aminobutyric acid (11.0 g/kg) may contribute to functional properties. The fatty acid composition was dominated by oleic acid (48.20%) and linoleic acid (29.20%), with minor contributions from saturated fatty acids such as palmitic and stearic acid. Detailed comparative data on the chemical composition and amino acid profile of the three *Y. lipolytica* strains (JII1a, PII6b and JII1c) are provided in the Supplementary Materials (Table S1).

The protein quality indices indicated that the biomass exhibited high nutritional adequacy, with a chemical score (CS) of 37.80% and an essential amino acid index (EAAI) of 36.17%, calculated relative to whole egg protein. A thorough microbiological analysis was conducted to ascertain the safety of the product. The analysis yielded low levels of aerobic microorganisms, negligible yeast and mould counts, and an absence of coliform bacteria and *Salmonella* spp.

Table 1. Nutritional composition characteristic of dried biomass from *Yarrowia lipolytica* JII1c strain.

	Nutritional Analysis	
Nutrient	Unit	Value [Mean ± SD]
Energy value	kJ (kcal)/100 g	1588 (375)
Dry matter	%	95.11 ± 0.95
Protein	%	43.12 ± 0.43
Carbohydrates, including:	%	32.34 ± 0.32
dietary fibres	%	32.32 ± 0.32
sugars	%	<0.20
Ash	%	11.00 ± 0.11
Fats, including:	%	7.03 ± 0.07
saturated fatty acids	%	0.50 ± 0.01
monounsaturated fatty acids	%	4.05 ± 0.04
polyunsaturated fatty acids	%	3.30 ± 0.03
Water content (moisture)	%	4.89 ± 0.05
Salt content	g/100 g	4.62 ± 0.05
Vitamins	Unit	Value [mean ± SD]
Vitamin B1 (thiamine)	mg/100 g	0.01 ± 0.01
Vitamin B2 (riboflavin)	mg/100 g	3.19 ± 0.14
Vitamin B6 (pyridoxine)	mg/100 g	0.20 ± 0.01
Vitamin B7 (biotin)	µg/100 g	205.00 ± 3.34
Vitamin B9 (folic acid)	µg/100 g	177.00 ± 7.06
Vitamin B12 (cyanocobalamin)	µg/100 g	0.33 ± 0.02
Macro and Microelements	Unit	Value [mean ± SD]
Calcium (Ca)	mg/100 g	27.50 ± 1.49
Phosphorus (P)	mg/100 g	2640.00 ± 13.94
Sodium (Na)	mg/100 g	1848 ± 76.57
Iron (Fe)	mg/100 g	24.00 ± 2.30
Copper (Cu)	mg/100 g	1.90 ± 0.08
Magnesium (Mg)	mg/100 g	138.00 ± 6.12
Potassium (K)	mg/100 g	2665 ± 15.33
Manganese (Mn)	mg/100 g	1.06 ± 0.04
Zinc (Zn)	mg/100 g	22.00 ± 0.77
Chromium (Cr)	mg/100 g	0.10 ± 0.01
Amino Acid Profile	Unit	Value [mean ± SD]
Aspartic acid	g/kg	41.10 ± 0.41
Glutamic acid	g/kg	56.00 ± 0.56
Arginine	g/kg	21.40 ± 0.21
Serine	g/kg	27.70 ± 0.28
Alanine	g/kg	31.10 ± 0.31
Glycine	g/kg	17.20 ± 0.17
Proline	g/kg	16.80 ± 0.17
Lysine	g/kg	31.10 ± 0.31
Methionine + Cystine	g/kg	9.30 ± 0.05

Table 1. *Cont.*

Nutritional Analysis		
Phenylalanine + Tyrosine	g/kg	35.20 ± 0.35
Threonine	g/kg	25.40 ± 0.25
Tryptophan	g/kg	5.10 ± 0.05
Leucine	g/kg	31.20 ± 0.31
Isoleucine	g/kg	23.60 ± 0.24
Valine	g/kg	25.10 ± 0.25
Histidine	g/kg	8.70 ± 0.09
Ornithine	g/kg	1.10 ± 0.03
Gamma-aminobutyric acid	g/kg	11.00 ± 0.61
Taurine	g/kg	< 0.05
Fatty Acids	Unit	Value [mean ± SD]
Butyric acid (C4:0)	%	<0.05
Myristic acid (C14:0)	%	<0.05
Pentadecanoic acid (C15:0)	%	0.48 ± 0.02
Palmitic acid (C16:0)	%	6.12 ± 0.20
Palmitoleic acid (C16:1)	%	0.38 ± 0.02
Heptadecanoic acid (C17:0)	%	0.19 ± 0.01
Stearic acid (C18:0)	%	1.05 ± 0.04
Oleic acid (C18:1, n9c)	%	48.20 ± 0.87
Linoelaidic acid (C18:2, n6t)	%	<0.05
Linoleic acid (C18:2, n6c)	%	29.20 ± 1.11
Arachidic acid (C20:0)	%	0.19 ± 0.01
cis-11.14-Eicosadienoic acid (C20:2)	%	<0.05
Behenic acid (C22:0)	%	<0.05
cis-11,17,17-Eicosatrienoic acid (C20:3, n3)	%	<0.05
Lignoceric acid (C24:0)	%	<0.05
Nutritional Indexes		
Parameter	Unit	Value [mean ± SD]
CS	%	37.80 ± 0.38
EAAI	%	36.17 ± 0.36

SD—standard deviation; CS—chemical score [%]; EAAI—essential amino acid index [%].

3.2. Protein Hydrolysis and Peptide Characterization

The progression of protein hydrolysis (Figure 2) was evaluated over a 24 h period using two enzymatic treatments: pepsin and a serine protease derived from figleaf pumpkin (*Cucurbita ficifolia*). As shown in Figure 2, minimal changes in DH values were observed in the control sample, ranging from 3.12% at 0 h to 3.41% at 24 h, indicating spontaneous, non-enzymatic breakdown of peptides [62]. A higher DH indicates more extensive protein breakdown, which can correlate with increased bioactivity and improved techno-functional properties of the hydrolysates.

The hydrolysis level statistically increased after enzymatic treatment with pepsin, reaching 11.72% after 24 h, with a clear stepwise progression at 5 h (6.42%) and 12 h (9.13%) (0 h vs. 24 h, $p = 0.000077$). Moreover, it is notable that the hydrolysis catalyzed by the figleaf protease yielded a statistically higher degree of hydrolysis (DH) over the course of the experiment, increasing from 3.16% at 0 h to 24.15% at 24 h (0 h vs. 24 h, $p = 0.000077$). The most pronounced increase occurred within the first 5 h (13.45%), indicating strong and rapid proteolytic activity.

Hydrolysate produced using figleaf protease exhibited the highest degree of hydrolysis, confirming the superior efficiency of this plant-derived enzyme in degrading *Y. lipolytica* protein. Consequently, this hydrolysate was selected for further investigation, including the assessment of its biological activities and techno-functional properties. The 24 h hydrolysate was selected for further analysis, including assessment of its biological activities and techno-functional properties.

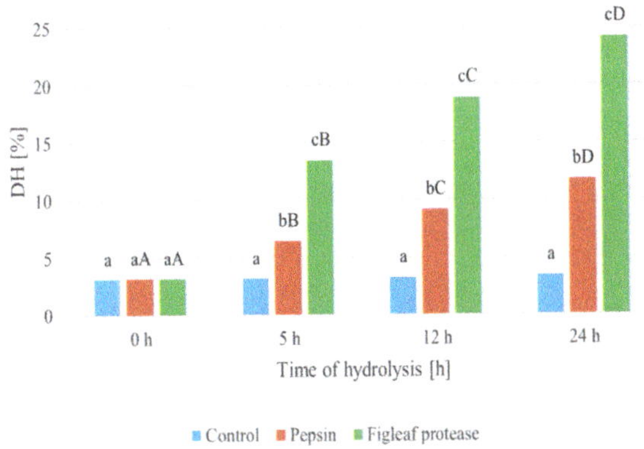

Figure 2. Degree of hydrolysis (DH [%]) during enzymatic degradation of protein extracted from *Y. lipolytica* JII1c yeast biomass, carried out using pepsin and serine protease from figleaf pumpkin at different times of reaction. Small letters (a–c) indicate statistically significant differences ($p < 0.05$) between sampling times within the same enzyme treatment. Capital letters (A–D) indicate statistically significant differences ($p < 0.05$) between enzyme types (control, pepsin, figleaf protease) at the same time point.

RP-HPLC analysis (Figure 3) demonstrated distinct differences in peptide profiles between the untreated protein isolate and the 24 h hydrolysate generated with *Cucurbita ficifolia* serine protease. The hydrolysate exhibited a broader distribution of peaks and increased peak complexity, particularly between 10 and 20 min of retention time, reflecting extensive proteolysis and the release of low molecular weight peptides. These chromatographic results are indicative of a more heterogeneous mixture of peptides, likely with varying molecular weights and bioactivities, and confirm the potent proteolytic activity of the figleaf pumpkin enzyme, capable of producing diverse peptide fractions.

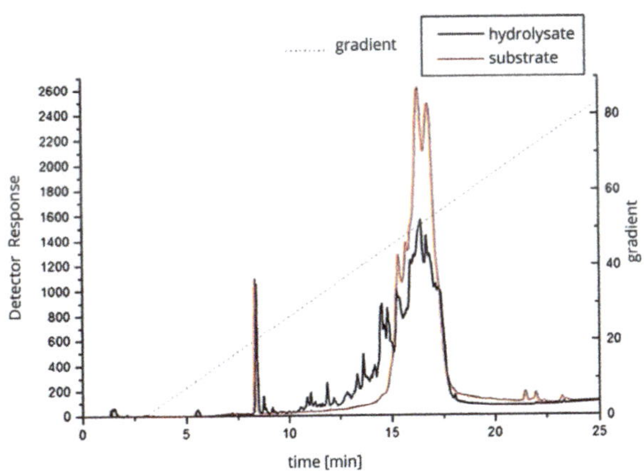

Figure 3. Protein–peptide separation of 24 h protein hydrolysate from *Y. lipolytica* JII1c yeast biomass and digested with serine protease from figleaf pumpkin.

3.3. Biological Activities of Hydrolysates

The 24 h enzymatic hydrolysate of *Yarrowia lipolytica* JII1c protein exhibited significantly enhanced biological activity in comparison with the native protein isolate (Table 2). The antioxidant activity, measured by DPPH and FRAP assays, increased more than twofold (from 0.390 to 0.87 μM Trolox/mg; $p = 0.000291$) and nearly tenfold (from 4.03 to 39.12 μg Fe^{3+}/mg; $p = 0.000191$), respectively. Concurrently, the Fe^{2+} ion chelating capacity exhibited a marked enhancement, rising from 275 to 1326 μg Fe^{2+}/mg ($p = 0.000272$).

Table 2. Level of biological activity in a 24 h hydrolysate solution of protein extracted from *Y. lipolytica* JII1c yeast biomass digested with serine protease from figleaf pumpkin.

Type of Biological Activity Determination Method	Sample	
	PI from *Y. lipolytica* JII1c Biomass	PH from *Y. lipolytica* JII1c Biomass
	Mean ± SD	Mean ± SD
Antioxidant activity (DPPH) [μM Trolox/mg]	0.39 ± 0.01 [a]	0.87 ± 0.00 [b]
Fe^{3+} ion reducing capacity (FRAP) [g Fe^{3+}/mg]	4.03 ± 0.01 [a]	39.12 ± 0.01 [b]
Fe^{2+} ion chelation capacity [μg Fe^{2+}/mg]	275.00 ± 0.58 [a]	1326.00 ± 0.58 [b]
Angiotensin-converting enzyme (ACE) inhibitory activity IC_{50} [mg/mL]	37.43 ± 0.06 [a]	8.20 ± 0.00 [b]
α-Glucosidase inhibitory activity IC_{50} [mg/mL]	12.23 ± 0.12 [a]	4.17 ± 0.06 [b]
DPP-IV inhibitory activity IC_{50} [mg/mL]	10.20 ± 0.00 [a]	2.40 ± 0.10 [b]

PI—protein isolate; PH—24 h protein hydrolysate; mean values (PI vs. PH) with different letters (a, b) indicate statistically different values ($p < 0.05$).

In terms of enzyme inhibition, the hydrolysate demonstrated a substantial decrease in IC_{50} values, indicating enhanced inhibitory potency. ACE inhibitory activity improved nearly fivefold (from 37.43 to 8.20 mg/mL; $p = 0.000291$). Additionally, α-glucosidase and DPP-IV inhibitory activities increased approximately three- to fourfold, with IC_{50} values dropping from 12.23 to 4.17 mg/mL ($p = 0.000261$) and from 10.20 to 2.40 mg/mL ($p = 0.000181$), respectively.

3.4. Functional Properties of Protein Preparations

The water absorption capacity of the hydrolysate more than doubled compared to the native protein (from 4.34 g H_2O/g to 9.12 g H_2O/g; $p = 0.000291$) (Table 3). This indicates an improved exposure of hydrophilic groups and a higher degree of molecular unfolding after hydrolysis. This facilitates water binding, which is advantageous for plant-based formulations requiring moisture retention.

A similar trend was observed in the OAC, which increased from 2.35 g oil/g in the protein isolate to 7.23 g oil/g in the hydrolysate ($p = 0.000391$). The elevated OAC may be attributed to the generation of smaller peptides with exposed nonpolar residues that interact more effectively with lipids, thereby enhancing the emulsion or mouthfeel properties in fat-containing food matrices.

The analysis of the NSI, which reflects protein dispersibility and processing potential, revealed more than a twofold increase in the hydrolysate compared to the isolate (49.20% vs. 19.40%; $p = 0.000291$).

Table 3. Functional properties of the protein isolate extracted from *Y. lipolytica* JII1c yeast biomass and its hydrolysate.

Type of Functional Property	Sample	
	PI from *Y. lipolytica* JII1c Biomass	**PH from *Y. Lipolytica* JII1c Biomass**
	Mean ± SD	**Mean ± SD**
Water absorption capacity (WAC) [g H_2O/g]	4.34 ± 0.04 [a]	9.12 ± 0.09 [b]
Oil absorption capacity (OAC) [g oil/g]	2.35 ± 0.02 [a]	7.23 ± 0.07 [b]
Nitrogen solubility index (NSI) [%]	19.40 ± 0.19 [a]	49.20 ± 0.49 [b]

PI—protein isolate; PH—24 h protein hydrolysate; mean values (PI vs. PH) with different letters (a, b) indicate statistically different values ($p < 0.05$).

This enhancement can be attributed to the breakdown of high-molecular-weight proteins into smaller, more soluble peptide fragments, thereby improving the protein's suitability for use in liquid or semi-liquid food systems.

The stability of egg white foam was found to be significantly influenced by the addition of yeast protein preparations from *Yarrowia lipolytica* JII1c biomass (Table 4). The incorporation of the protein isolate resulted in a notable enhancement of foam persistence, which was observed to be dose-dependent. At the highest concentration of yeast PI (3%), the leakage volume was reduced to 0 mL (100% of retained foam volume), indicating full foam stability, in contrast to 29.0 mL in the control sample without any additive ($p = 0.000059$). Lower doses of yeast protein isolate (1% and 2%) also demonstrated a statistically significant improvement in foam stability, in comparison to the control sample, with leakage reduced to 6.6 mL (77.2% of retained foam volume) and 1.9 mL (93.5% of retained foam volume), respectively.

In contrast, the 24 h hydrolysate from *Y. lipolytica* JII1c showed much lower effectiveness in stabilising the foam (Table 4). Even at the highest tested concentration of 3%, the leakage volume was 2.4 mL (91.7% of retained foam volume), while at 1% and 2% doses, the leakage remained relatively high at 28.0 mL (3.5% of retained foam volume) and 6.8 mL (76.6% of retained foam volume) ($p = 0.000152$), respectively.

The addition of yeast protein preparations from *Yarrowia lipolytica* JII1c resulted in a significant enhancement of emulsion stability in a model oil-in-water system (1:1 ratio). The control sample exhibited a gradual decline in stability over time, dropping from 64.87% at 0.5 h to 53.36% at 3 h. Both the protein isolate and its 24 h hydrolysate demonstrated markedly higher and more sustained emulsion stability.

For the isolate, emulsion stability improved with increasing dose. At 3%, emulsion stability remained above 95% throughout the 3 h period, reaching 97.41% at 0.5 h and 95.11% at 3 h. A similar trend was observed for the 1% and 2% doses, though with slightly lower stability than the highest dose. A notable observation was the superior performance of the hydrolysate in comparison to the isolate, particularly in the initial stages of the test. All concentrations of hydrolysate tested yielded consistent and high stability values, exceeding 96% throughout the entire testing period. The highest result was noted for the 3% dose at 0.5 h (96.86%) and 3 h (95.75%), indicating superior emulsifying properties.

Table 4. Persistence of egg white foam and emulsion stability of yeast protein isolate *Y. lipolitica* JII1c and its hydrolysate.

Type of Substance	Dose [%]	Leakage Volume After 30 min [mL]	Emulsion Stability [%]			
			0.5 h	1 h	2 h	3 h
		Mean ± SD	Mean ± SD	Mean ± SD	Mean ± SD	Mean ± SD
Control	-	29.0 ± 0.3 [a]	64.87 ± 0.48 [a]	56.85 ± 0.48 [a]	54.07 ± 0.69 [a]	53.36 ± 0.6
PI from *Y. lipolytica* JII1c biomass	1	6.6 ± 0.1 [b,A]	90.80 ± 0.72 [b]	90.49 ± 0.99 [b]	84.17 ± 0.67 [b]	83.44 ± 1.32 [b]
	2	1.9 ± 0.02 [c,A]	96.80 ± 0.89 [b]	95.96 ± 1.22 [b]	95.60 ± 0.53 [b]	92.92 ± 0.67 [b]
	3	0.0 ± 0.0 [d,A]	97.41 ± 1.41 [b]	96.49 ± 1.43 [b]	95.74 ± 0.20 [b]	95.11 ± 0.23 [b]
PH from *Y. lipolytica* JII1c biomass	1	28.0 ± 0.3 [a,B]	96.51 ± 1.22 [b]	96.53 ± 0.59 [b]	95.88 ± 2.75 [b]	95.31 ± 2.01 [b]
	2	6.8 ± 0.1 [e,B]	96.68 ± 1.04 [b]	96.47 ± 1.37 [b]	96.36 ± 0.72 [b]	95.42 ± 0.54 [b]
	3	2.4 ± 0.02 [f,B]	96.86 ± 1.30 [b]	96.66 ± 1.05 [b]	96.39 ± 0.41 [b]	95.75 ± 1.40 [b]

PI—protein isolate; PH—24 h protein hydrolysate; mean ± SD—average value ± standard deviation; small letters ([a, b, c] etc.) indicate statistically significant values ($p < 0.05$) for leakage: control vs. PI (1, 2, 3%) and control vs. PH (1, 2, 3%); big letters ([A, B]) indicate statistically significant values ($p < 0.05$) between PI 1% vs. PH 1%, PI 2% vs. PH 2%, and PI 3% vs. PH 3%; small letters ([a, b]) indicate statistically significant values ($p < 0.05$) for emulsion stability: control (0.5, 1, 2, 3 h) vs. PI (1, 2, 3%) and control (0.5, 1, 2, 3 h) vs. PH (1, 2, 3%).

3.5. Sensory Evaluation of Extrudates

The organoleptic analysis of extrudates supplemented with yeast protein preparations from *Yarrowia lipolytica* JII1c revealed a clear preference for the hydrolysate over the isolate across most sensory attributes (Figure 4). The hydrolysate-enriched extrudates obtained higher scores in terms of taste (4.5 and 3.5), smell (4.0 and 3.0), texture (4.25 and 2.75), and appearance (4.25 and 3.25). Notably, saltiness perception was consistently unaltered for both samples at a rating of 3.5, indicating that the process of enzymatic hydrolysis did not affect the perception of saltiness.

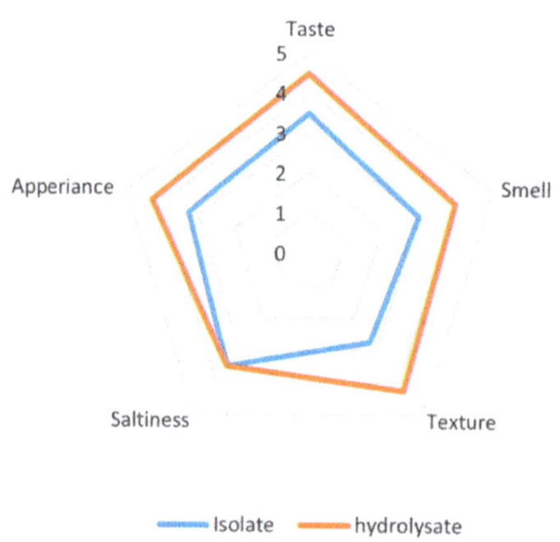

Figure 4. Organoleptic evaluation of extrudates prepared from a protein isolate extracted from *Y. lipolytica* JII1c yeast biomass and its hydrolysate.

The most significant differences were observed in texture and aroma, suggesting that the hydrolysate contributed to more appealing sensory properties, such as mouthfeel and aroma profile, possibly due to the formation of flavour-active peptides during hydrolysis. These findings underscore the potential of protein hydrolysates from *Y. lipolytica* to enhance the sensory profile of extruded food products.

4. Discussion

The primary challenge in the utilisation of yeasts in food applications lies in the efficient release of intracellular components through cell wall disintegration. Another crucial aspect is the conversion of proteins into bioactive peptides while maintaining low processing costs. The findings of this study demonstrate that strain selection is a pivotal factor in determining techno-economic feasibility. The findings of this study demonstrate that strain selection is a pivotal factor influencing process efficiency and the potential applicability of yeast-derived proteins in food production [9,63–66]. Such gains directly support sustainability goals by reducing substrate input and enabling valorisation of industrial side streams.

The nutritional profile of JII1c, characterized by its high protein content and balanced essential amino acid (EAA) pattern, supports its role as an alternative to conventional proteins. This profile is consistent with the ranges reported for *Y. lipolytica* across various substrates and strains [16,67–69]. Although sulphur-containing amino acids continue to be a limiting factor, comprehensive protein quality indices suggest that the EAA composition is appropriate for incorporation into plant-based formulations, particularly due to its ability to complement cereal proteins, as reported by Michalik et al. (2014) [68]. In addition, the presence of dietary fibre and low residual sugars is conducive to a low glycaemic design, while the micronutrient package (iron, zinc, B vitamins including B12) contributes nutritional value relevant to vegan diets and reflects the species' metabolic flexibility on biofuel-derived streams [25,70]. These characteristics are consistent with the regulatory framework of *Y. lipolytica* biomass as a novel food supplement in the EU, which stipulates explicit utilisation thresholds [12,71].

With regard to the processing strategy, enzymatic hydrolysis remains the most effective method of solubilising yeast matrices and tailoring protein functionality through controlled cleavage, exposure of reactive side chains, and modulation of molecular size without damaging the proteins [72,73]. In the present study, a plant-derived serine protease from *Cucurbita ficifolia* was found to exhibit superior proteolytic efficacy in comparison to pepsin, as evidenced by the distinct peptide distributions observed by RP-HPLC. This finding aligns with the current understanding that the selection of enzymes and the prevailing conditions of a reaction (time, pH, enzyme-to-substrate ratio) regulate the ratio of peptides to free amino acids, thereby influencing the balance between functionality and taste [74–77]. Elevated levels of hydrolysis can lead to an increase in free amino acids, accompanied by a reduction in longer peptides. Consequently, the optimisation process should be oriented towards application-specific endpoints rather than the pursuit of maximal DH alone [75].

Functionally, hydrolysis enhanced water and oil binding and markedly increased solubility, properties critical for moist textures and dispersibility in beverages or semi-liquid systems. Mechanistically, peptide fragmentation and unfolding expose hydrophilic and hydrophobic groups, thereby enabling improved interactions with both aqueous and lipid phases [78–81]. The observed trade-off between foaming and emulsion stability, with the former being reduced and the latter enhanced, aligns with interfacial principles. This is evidenced by the rapid migration of smaller peptides, which can form cohesive interfacial films that stabilise oil-in-water emulsions. A similar phenomenon has been reported for yeast-derived bioemulsifiers [81].

Beyond functionality, the bioactivities of the hydrolysate were substantially higher than those of the native isolate. This is in line with literature on yeast-derived antioxidant and enzyme inhibitory peptides relevant to cardiometabolic health [5,82–84]. Previous studies have demonstrated that the hydrolysate exhibits immunomodulatory, antimicrobial, and anti-inflammatory properties, indicating potential for broader health benefits [85–87]. The sensory improvements observed in extrudates are consistent with the enhanced release of amino nitrogen and small taste-active peptides contributing to umami, mouthfulness, and flavour continuity [88]. This relationship between the biochemical activities and sensory perception suggests that peptides responsible for antioxidant and enzyme inhibitory effects may also modulate taste responses, thereby linking nutritional and sensory functionality. Recent findings further indicate that yeast peptides may enhance saltiness perception through receptor interactions, offering a plausible method for reducing sodium levels without compromising flavour intensity [89].

The following section will address the practical implications of the aforementioned points. For manufacturers, *Y. lipolytica* JII1c provides a flexible platform for the generation of application-tuned peptide fractions. The benefits that can be derived from these emulsions and beverages include high solubility and emulsion stability. Meat analogues and baked systems, meanwhile, leverage water/oil binding. Finally, savoury snacks and extruded matrices gain from umami-active peptides and potential salt reduction. The utilisation of a plant protease has been demonstrated to enhance consumer acceptance of processing aids, thereby aligning with prevailing sustainability narratives.

Although the current results highlight the potential of *Y. lipolytica* proteins, there are still several challenges to overcome. The bioavailability and stability of the proteins during processing and storage, as well as their sensory thresholds in complex foods, require confirmation in both model and real systems. Future work should also address bitterness at higher degrees of hydrolysis and identify bioactive peptides. Finally, techno-economic analyses are required to benchmark enzyme costs and scaling-up scenarios.

5. Conclusions

Yarrowia lipolytica JII1c exhibited elevated levels of protein and significant biomass yield, in addition to high levels of dietary fibre and essential micronutrients (e.g., iron, zinc, B vitamins), indicating its potential as an alternative protein source.

The amino acid profile of the biomass demonstrated a balanced EAA composition, and although sulphur-containing amino acids remained limiting, the overall protein quality (CS and EAAI) supports its use in plant-based nutrition systems.

Enzymatic hydrolysis using serine protease from *Cucurbita ficifolia* markedly enhanced several techno-functional properties of the protein isolates, notably water absorption capacity, oil absorption capacity and nitrogen solubility, which are key attributes for developing novel food formulations. The process also generated bioactive peptides with increased antioxidant and enzyme inhibitory activities that, in addition to their physiological potential, enhanced the perception of umami and saltiness, demonstrating their promise as natural flavour enhancers for reduced-sodium and clean-label food products.

The study highlights the multifunctional character of *Y. lipolytica*, combining nutritional richness, functional versatility, and sensory benefits, which strongly supports its use as a natural, health-promoting ingredient in sustainable food production. The yeast's natural origin, efficient bioprocess scalability, and ability to valorise agro-industrial by-products further strengthen its relevance to the circular bioeconomy model and to the growing demand for environmentally responsible protein sources.

In order to apply these results in a practical context, it is essential that future research focus on evaluating the bioavailability of the released peptides, their stability during food processing and storage, and comprehensive toxicological and safety assessments for human consumption. Furthermore, the investigation of sensory optimisation, digestive stability, and techno-economic feasibility under industrial conditions will provide a clearer path towards the commercial utilisation of *Y. lipolytica*-derived proteins and peptides in modern nutrition.

Supplementary Materials: The following supporting information can be downloaded at: https://www.mdpi.com/article/10.3390/foods14213801/s1, Table S1: Nutritional composition characteristic of dried biomass from *Yarrowia lipolytica* JII1a, JII1c, an PII6b strains.

Author Contributions: Conceptualization, M.S.; methodology, M.S. and A.D.; software, A.M.; validation, M.S.; formal analysis, M.M.; investigation, M.M.; resources, M.S.; data curation, M.M.; writing—original draft preparation, M.N.; writing—review and editing, A.D.; visualization, A.M.; supervision, A.D.; project administration, M.S.; funding acquisition, M.S. All authors have read and agreed to the published version of the manuscript.

Funding: This work was supported by the Wrocław University of Environmental and Life Sciences (Poland), the Department of Research Administration (DRA) as part of the research project "The use of *Yarrowia lipolytica* yeast protein as a potential source of biologically active peptides and an ingredient useful in food processing"/no. N090/0011/23.

Institutional Review Board Statement: The study was conducted in accordance with the Declaration of Helsinki, and approved by the Institutional Review Board (or Ethics Committee) of the Rector's Committee for Ethics in Scientific Research (No. N0N00000.020.1.8.4.2024, 21 October 2024).

Informed Consent Statement: Informed consent was obtained from all subjects involved in the study.

Data Availability Statement: The original contributions presented in this study are included in the article/Supplementary Materials. Further inquiries can be directed to the corresponding author.

Conflicts of Interest: The authors declare no conflicts of interest.

References

1. Gorissen, S.H.M.; Witard, O.C. Characterising the Muscle Anabolic Potential of Dairy, Meat and Plant-Based Protein Sources in Older Adults. *Proc. Nutr. Soc.* **2018**, *77*, 20–31. [CrossRef] [PubMed]
2. Sobczak, P.; Grochowicz, J.; Łusiak, P.; Żukiewicz-Sobczak, W. Development of Alternative Protein Sources in Terms of a Sustainable System. *Sustainability* **2023**, *15*, 12111. [CrossRef]
3. Berners-Lee, M.; Kennelly, C.; Watson, R.; Hewitt, C.N. Current Global Food Production Is Sufficient to Meet Human Nutritional Needs in 2050 Provided There Is Radical Societal Adaptation. *Elem. Sci. Anthr.* **2018**, *6*, 52. [CrossRef]
4. Li, X.; Cao, Q.; Liu, G. Advances, Applications, Challenges and Prospects of Alternative Proteins. *J. Food Compos. Anal.* **2025**, *137*, 106900. [CrossRef]
5. Ma, J.; Sun, Y.; Meng, D.; Zhou, Z.; Zhang, Y.; Yang, R. Yeast Proteins: The Novel and Sustainable Alternative Protein in Food Applications. *Trends Food Sci. Technol.* **2023**, *135*, 190–201. [CrossRef]
6. Aschemann-Witzel, J.; Gantriis, R.F.; Fraga, P.; Perez-Cueto, F.J.A. Plant-Based Food and Protein Trend from a Business Perspective: Markets, Consumers, and the Challenges and Opportunities in the Future. *Crit. Rev. Food Sci. Nutr.* **2021**, *61*, 3119–3128. [CrossRef]
7. Abe-Inge, V.; Aidoo, R.; Moncada de la Fuente, M.; Kwofie, E.M. Plant-Based Dietary Shift: Current Trends, Barriers, and Carriers. *Trends Food Sci. Technol.* **2024**, *143*, 104292. [CrossRef]
8. Fatima, N.; Emambux, M.N.; Olaimat, A.N.; Stratakos, A.C.; Nawaz, A.; Wahyono, A.; Gul, K.; Park, J.; Shahbaz, H.M. Recent Advances in Microalgae, Insects, and Cultured Meat as Sustainable Alternative Protein Sources. *Food Humanit.* **2023**, *1*, 731–741. [CrossRef]
9. Wang, B.; Shi, Y.; Lu, H.; Chen, Q. A Critical Review of Fungal Proteins: Emerging Preparation Technology, Active Efficacy and Food Application. *Trends Food Sci. Technol.* **2023**, *141*, 104178. [CrossRef]
10. Shanmugam, S.R.; Schorer, R.; Arthur, W.; Drabold, E.; Rudar, M.; Higgins, B. Upcycling Nutrients from Poultry Slaughterhouse Wastewater through Cultivation of the Nutritional Yeast, *Yarrowia lipolytica*. *J. Environ. Chem. Eng.* **2025**, *13*, 115245. [CrossRef]
11. Cao, X.; Liu, H.; Yang, M.; Mao, K.; Wang, X.; Chen, Z.; Ran, M.; Hao, L. Evaluation of the Nutritional Quality of Yeast Protein in Comparison to Animal and Plant Proteins Using Growing Rats and INFOGEST Model. *Food Chem.* **2025**, *463*, 141178. [CrossRef]
12. Turck, D.; Castenmiller, J.; de Henauw, S.; Hirsch-Ernst, K.; Kearney, J.; Maciuk, A.; Mangelsdorf, I.; McArdle, H.J.; Naska, A.; Pelaez, C.; et al. Safety of *Yarrowia lipolytica* Yeast Biomass as a Novel Food Pursuant to Regulation (EU) 2015/2283. *EFSA J.* **2019**, *17*, e05594. [CrossRef] [PubMed]
13. Mamaev, D.; Zvyagilskaya, R. *Yarrowia lipolytica*: A Multitalented Yeast Species of Ecological Significance. *FEMS Yeast Res.* **2021**, *21*, foab008. [CrossRef] [PubMed]

14. Gottardi, D.; Ciccone, M.; Siroli, L.; Lanciotti, R.; Patrignani, F. Use of *Yarrowia lipolytica* to Obtain Fish Waste Functional Hydrolysates Rich in Flavoring Compounds. *Fermentation* **2022**, *8*, 708. [CrossRef]
15. Carsanba, E.; Agirman, B.; Papanikolaou, S.; Fickers, P.; Erten, H. Valorisation of Waste Bread for the Production of Yeast Biomass by *Yarrowia lipolytica* Bioreactor Fermentation. *Fermentation* **2023**, *9*, 687. [CrossRef]
16. Drzymała, K.; Mirończuk, A.M.; Pietrzak, W.; Dobrowolski, A. Rye and Oat Agricultural Wastes as Substrate Candidates for Biomass Production of the Non-Conventional Yeast *Yarrowia lipolytica*. *Sustainability* **2020**, *12*, 7704. [CrossRef]
17. Lalić, A.; Jagelavičiūtė, J.; Rezić, T.; Trivunović, Z.; Žadeikė, D.; Bašinskienė, L. From Bakery Leftovers to Brewing Sustainability: Fermentation of Spent Grain with *Yarrowia lipolytica* and Lactobacillus Acidophilus. *Sustainability* **2025**, *17*, 782. [CrossRef]
18. Chakrabarti, S.; Guha, S.; Majumder, K. Food-Derived Bioactive Peptides in Human Health: Challenges and Opportunities. *Nutrients* **2018**, *10*, 1738. [CrossRef]
19. Mazorra-Manzano, M.A.; Mora-Cortes, W.G.; Leandro-Roldan, M.M.; González-Velázquez, D.A.; Torres-Llanez, M.J.; Ramírez-Suarez, J.C.; González-Córdova, A.F.; Vallejo-Córdoba, B. Production of Whey Protein Hydrolysates with Angiotensin-Converting Enzyme-Inhibitory Activity Using Three New Sources of Plant Proteases. *Biocatal. Agric. Biotechnol.* **2020**, *28*, 101724. [CrossRef]
20. Dąbrowska, A.; Szołtysik, M.; Babij, K.; Pokora, M.; Zambrowicz, A.; Chrzanowska, J. Application of Asian Pumpkin (*Cucurbita ficifolia*) Serine Proteinase for Production of Biologically Active Peptides from Casein. *Acta Biochim. Pol.* **2013**, *60*, 117–122. [CrossRef]
21. Maciejewska, M.; Dąbrowska, A.; Cano-Lamadrid, M. Sustainable Protein Sources: Functional Analysis of Tenebrio Molitor Hydrolysates and Attitudes of Consumers in Poland and Spain Toward Insect-Based Foods. *Foods* **2025**, *14*, 333. [CrossRef] [PubMed]
22. Zambrowicz, A.; Eckert, E.; Pokora, M.; Bobak, Ł.; Dąbrowska, A.; Szołtysik, M.; Trziszka, T.; Chrzanowska, J. Antioxidant and Antidiabetic Activities of Peptides Isolated from a Hydrolysate of an Egg-Yolk Protein by-Product Prepared with a Proteinase from Asian Pumpkin (*Cucurbita ficifolia*). *RSC Adv* **2015**, *5*, 10460–10467. [CrossRef]
23. Babij, K.; Bajzert, J.; Dąbrowska, A.; Szołtysik, M.; Zambrowicz, A.; Lubec, G.; Stefaniak, T.; Willak-Janc, E.; Chrzanowska, J. Hydrolysis with *Cucurbita ficifolia* Serine Protease Reduces Antigenic Response to Bovine Whey Protein Concentrate and As-Casein. *Amino Acids* **2015**, *47*, 2335–2343. [CrossRef] [PubMed]
24. Da Silva, R.M.; da Silva, C.N.; de Faria-Júnior, C.S.; Buarque, F.S.; Ribeiro, B.D.; Lemes, A.C.; Coelho, M.A.Z. Extraction and Characterization of High-Value Compounds from *Yarrowia lipolytica* W29 Using Sequential Hydrolysis. *Processes* **2025**, *13*, 615. [CrossRef]
25. Jach, M.E.; Malm, A. *Yarrowia lipolytica* as an Alternative and Valuable Source of Nutritional and Bioactive Compounds for Humans. *Molecules* **2022**, *27*, 2300. [CrossRef]
26. AOAC International. *Official Method of Analysis for Dry Matter*, 18th ed.; AOAC International: Arlington, TX, USA, 2007.
27. AOAC International. *AOAC Official Method 2011.25: Insoluble, Soluble and Total Dietary Fiber in Foods. Enzymatic-Gravimetric-Liquid Chromatography*; AOAC International: Arlington, TX, USA, 2011.
28. Masuko, T.; Minami, A.; Iwasaki, N.; Majima, T.; Nishimura, S.-I.; Lee, Y.C. Carbohydrate Analysis by a Phenol–Sulfuric Acid Method in Microplate Format. *Anal. Biochem.* **2005**, *339*, 69–72. [CrossRef]
29. AOAC International. *Method 996.06, Fat (Total, Saturated, and Monounsaturated) in Foods. Official Methods of Analysis*; AOAC International: Arlington, TX, USA, 1995.
30. Bremner, J.M. Determination of Nitrogen in Soil by the Kjeldahl Method. *J. Agric. Sci.* **1960**, *55*, 11–33. [CrossRef]
31. *EN 14084:2003*; Foodstuffs—Determination of Trace Elements—Determination of Calcium, Copper, Iron, Magnesium, Manganese, Sodium, Phosphorus, Potassium and Zinc by ICP-OES. European Committee for Standardization: Brussels, Belgium, 2003.
32. López-Bascón, M.A.; Luque de Castro, M.D. Soxhlet Extraction. In *Liquid-Phase Extraction*; Elsevier: Amsterdam, The Netherlands, 2020; pp. 327–354.
33. AOAC. *Official Method 944.12 Folic Acid in Vitamin Preparation, Microbiological Methods*; AOAC: Gaithersburg, MD, USA, 1996.
34. AOAC. *Official Method 992.05 Folic Acid in Infant Formula, Microbiological Methods*; AOAC: Gaithersburg, MD, USA, 2016.
35. AOAC. *985.32 Vitamin B6 (Pyridoxine, Pyridoxal, Pyridoxamine) in Ready-to-Feed Milk Based Infant Formula*; AOAC: Gaithersburg, MD, USA, 1988.
36. AOAC. *961.15 Vitamin B6 (Pyridoxine, Pyridoxal, Pyridoxamine) in Food Extracts, Microbiological Methods*; AOAC: Gaithersburg, MD, USA, 1975.
37. AOAC. *Official Method 960.46. Vitamin Assays, Microbiological Method*; AOAC: Arlington, TX, USA, 2006.
38. AOAC. *Official Method 940.33 Riboflavin (Vitamin B2) in Vitamin Preparations, Microbiological Methods*; AOAC: Gaithersburg, MD, USA, 2006.
39. *PN-EN 13805:2003*; Foodstuffs. Determination of Trace Elements. Pressure Mineralization. Polish Committee for Standardization: Warsaw, Poland, 2004.
40. Ardö, Y.; Gripon, J.-C. Comparative Study of Peptidolysis in Some Semi-Hard Round-Eyed Cheese Varieties with Different Fat Contents. *J. Dairy Res.* **1995**, *62*, 543–547. [CrossRef]

41. Liang, X.; Qian, G.; Yang, H.; Chen, N.; Ai, Z.; Xing, Y.; Huang, W.; Xu, L.; Li, M.; Wang, Z.; et al. Evaluation of IgG/IgE-binding Capacity and Functional Properties of Enzymatic Hydrolysis in Skimmed Cow Milk System. *J. Food Sci.* **2023**, *88*, 2780–2795. [CrossRef]
42. WHO. Energy and Protein Requirements. In *Report of a Joint FAO/WHO/UNU Expert Consultation*; World Health Organization: Geneva, Switzerland, 1985; Volume 724, pp. 1–206.
43. FAO. *Food Energy—Methods of Analysis and Conversion Factors; FAO Food and Nutrition Paper 77*; FAO: Rome, Italy, 2003.
44. *ISO 4833-1:2013*; Microbiology of the Food Chain—Horizontal Method for the Enumeration of Microorganisms—Part 1: Colony Count at 30 °C by the Pour Plate Technique. ISO: Geneva, Switzerland, 2013.
45. *ISO 21527-2:2008*; Microbiology of Food and Animal Feeding Stuffs—Horizontal Method for the Enumeration of Yeasts and Moulds—Part 2: Colony Count Technique in Products with Water Activity Less than or Equal to 0.95. ISO: Geneva, Switzerland, 2008.
46. *ISO 4832:2006*; Microbiology of Food and Animal Feeding Stuffs—Horizontal Method for the Enumeration of Coliforms—Colony Count Technique. ISO: Geneva, Switzerland, 2006.
47. *ISO 6579-1:2017*; Microbiology of the Food Chain—Horizontal Method for the Detection, Enumeration and Serotyping of Salmonella—Part 1: Detection of Salmonella spp. ISO: Geneva, Switzerland, 2017.
48. Shetty, K.J.; Kinsella, J.E. Preparation of Yeast Protein Isolate with Low Nucleic Acid by Succinylation. *J. Food Sci.* **1979**, *44*, 633–638. [CrossRef]
49. Dryjanski, M.; Otlewski, J.; Polanowski, A.; Wilusz, T. Serine Proteinase from *Cucurbita ficifolia* Seed; Purification, Properties, Substrate Specificity and Action on Native Squash Trypsin Inhibitor (CMTI I). *Biol. Chem. Hoppe Seyler* **1990**, *371*, 889–896. [CrossRef] [PubMed]
50. Silvestre, M.P.C. Review of Methods for the Analysis of Protein Hydrolysates. *Food Chem.* **1997**, *60*, 263–271. [CrossRef]
51. Yen, G.-C.; Chen, H.-Y. Antioxidant Activity of Various Tea Extracts in Relation to Their Antimutagenicity. *J. Agric. Food Chem.* **1995**, *43*, 27–32. [CrossRef]
52. Benzie, I.F.F.; Strain, J.J. The Ferric Reducing Ability of Plasma (FRAP) as a Measure of "Antioxidant Power": The FRAP Assay. *Anal. Biochem.* **1996**, *239*, 70–76. [CrossRef]
53. Xu, X.; Katayama, S.; Mine, Y. Antioxidant Activity of Triptic Digest of Hen Egg Yolk Phosvitin. *J. Sci. Food Agric.* **2007**, *87*, 2604–2608. [CrossRef]
54. Hernández-Ledesma, B.; Miralles, B.; Amigo, L.; Ramos, M.; Recio, I. Identification of Antioxidant and ACE-Inhibitory Peptides in Fermented Milk. *J. Sci. Food Agric.* **2005**, *85*, 1041–1048. [CrossRef]
55. Yu, Z.; Yin, Y.; Zhao, W.; Liu, J.; Chen, F. Anti-Diabetic Activity Peptides from Albumin against α-Glucosidase and α-Amylase. *Food Chem.* **2012**, *135*, 2078–2085. [CrossRef]
56. Tulipano, G.; Sibilia, V.; Caroli, A.M.; Cocchi, D. Whey Proteins as Source of Dipeptidyl Dipeptidase IV (Dipeptidyl Peptidase-4) Inhibitors. *Peptides* **2011**, *32*, 835–838. [CrossRef]
57. Timilsena, Y.P.; Adhikari, R.; Barrow, C.J.; Adhikari, B. Physicochemical and Functional Properties of Protein Isolate Produced from Australian Chia Seeds. *Food Chem.* **2016**, *212*, 648–656. [CrossRef]
58. Wu, H.; Wang, Q.; Ma, T.; Ren, J. Comparative Studies on the Functional Properties of Various Protein Concentrate Preparations of Peanut Protein. *Food Res. Int.* **2009**, *42*, 343–348. [CrossRef]
59. Achouri, A.; Nail, V.; Boye, J.I. Sesame Protein Isolate: Fractionation, Secondary Structure and Functional Properties. *Food Res. Int.* **2012**, *46*, 360–369. [CrossRef]
60. Rytel, E.; Kita, A.; Pęksa, A.; Tajner-Czopek, A.; Miedzianka, J. Wpływ Zastosowania Soli w Produkcji Chrupek Kukurydzianych Wzbogaconych Dodatkiem Niekonwencjonalnych Surowców Na Wybrane Cechy Jakościowe. *Bromat. Chem. Toksykol.* **2015**, *48*, 512–517.
61. *ISO 13299:2016*; Sensory Analysis—Methodology—General Guidance for Establishing a Sensory Profile. ISO: Geneva, Switzerland, 2016.
62. Butré, C.I.; Buhler, S.; Sforza, S.; Gruppen, H.; Wierenga, P.A. Spontaneous, Non-Enzymatic Breakdown of Peptides during Enzymatic Protein Hydrolysis. *Biochim. Biophys. Acta (BBA)-Proteins Proteom.* **2015**, *1854*, 987–994. [CrossRef] [PubMed]
63. Bruder, S.; Melcher, F.A.; Zoll, T.; Hackenschmidt, S.; Kabisch, J. Evaluation of a *Yarrowia lipolytica* Strain Collection for Its Lipid and Carotenoid Production Capabilities. *Eur. J. Lipid Sci. Technol.* **2020**, *122*, 1900172. [CrossRef]
64. Liu, L.; Pan, A.; Spofford, C.; Zhou, N.; Alper, H.S. An Evolutionary Metabolic Engineering Approach for Enhancing Lipogenesis in *Yarrowia lipolytica*. *Metab. Eng.* **2015**, *29*, 36–45. [CrossRef] [PubMed]
65. Tsirigka, A.; Theodosiou, E.; Patsios, S.I.; Tsoureki, A.; Andreadelli, A.; Papa, E.; Aggeli, A.; Karabelas, A.J.; Makris, A.M. Novel Evolved *Yarrowia lipolytica* Strains for Enhanced Growth and Lipid Content under High Concentrations of Crude Glycerol. *Microb. Cell Fact.* **2023**, *22*, 62. [CrossRef]
66. Wang, S.; Yang, Y.; Yu, K.; Xu, S.; Liu, M.; Sun, J.; Zheng, J.; Zhang, Y.; Yuan, W. Engineering of *Yarrowia lipolytica* for Producing Pyruvate from Glycerol. *3 Biotech.* **2022**, *12*, 98. [CrossRef]

67. Juszczyk, P.; Tomaszewska, L.; Kita, A.; Rymowicz, W. Biomass Production by Novel Strains of *Yarrowia lipolytica* Using Raw Glycerol, Derived from Biodiesel Production. *Bioresour. Technol.* **2013**, *137*, 124–131. [CrossRef]
68. Michalik, B.; Biel, W.; Lubowicki, R.; Jacyno, E. Chemical Composition and Biological Value of Proteins of the Yeast *Yarrowia lipolytica* Growing on Industrial Glycerol. *Can. J. Anim. Sci.* **2014**, *94*, 99–104. [CrossRef]
69. Patsios, S.I.; Dedousi, A.; Sossidou, E.N.; Zdragas, A. Sustainable Animal Feed Protein through the Cultivation of *Yarrowia lipolytica* on Agro-Industrial Wastes and by-Products. *Sustainability* **2020**, *12*, 1398. [CrossRef]
70. Jach, M.E.; Sajnaga, E.; Janeczko, M.; Juda, M.; Kochanowicz, E.; Baj, T.; Malm, A. Production of Enriched in B Vitamins Biomass of *Yarrowia lipolytica* Grown in Biofuel Waste. *Saudi J. Biol. Sci.* **2021**, *28*, 2925–2932. [CrossRef]
71. Timira, V.; Chen, X.; Zhou, P.; Wu, J.; Wang, T. Potential Use of Yeast Protein in Terms of Biorefinery, Functionality, and Sustainability in Food Industry. *Compr. Rev. Food Sci. Food Saf.* **2024**, *23*, e13326. [CrossRef]
72. Celus, I.; Brijs, K.; Delcour, J.A. Enzymatic Hydrolysis of Brewers' Spent Grain Proteins and Technofunctional Properties of the Resulting Hydrolysates. *J. Agric. Food Chem.* **2007**, *55*, 8703–8710. [CrossRef]
73. Oliveira, A.S.; Ferreira, C.; Pereira, J.O.; Pintado, M.E.; Carvalho, A.P. Valorisation of Protein-Rich Extracts from Spent Brewer's Yeast (*Saccharomyces cerevisiae*): An Overview. *Biomass Convers. Biorefin.* **2025**, *15*, 1771–1793. [CrossRef]
74. Della Rosa, F.; Tonin, A.; Rocha, B.; Santos, M.; Silveira, F.; Cardoso-Filho, L.; Ribeiro, V.; Meurer, E. Optimization of Hydrolysis and Identification of Bioactive Peptides in Brewery Yeast Residuals. *J. Braz. Chem. Soc.* **2024**, *35*, e20230146. [CrossRef]
75. Min, J.H.; Lee, Y.J.; Kang, H.J.; Moon, N.R.; Park, Y.K.; Joo, S.-T.; Jung, Y.H. Characterization of Yeast Protein Hydrolysate for Potential Application as a Feed Additive. *Food Sci. Anim. Resour.* **2024**, *44*, 723–737. [CrossRef] [PubMed]
76. Yi, D.; Lin, Q.; Johns, P.W. Estimation of Degree of Hydrolysis of Protein Hydrolysates by Size Exclusion Chromatography. *Food Anal. Methods* **2021**, *14*, 805–813. [CrossRef]
77. Suh, H.J.; Shin, J.C.; Kim, J.H.; Jang, J.H.; Han, S.H. Optimal Enzyme Selection for Organic Whey Protein Hydrolysis. *Korean Soc. Food Nutr.* **2017**, *30*, 1359–1363.
78. Dent, T.; LeMinh, A.; Maleky, F. Comparison of Colorimetric Methods for Measuring the Solubility of Legume Proteins. *Gels* **2024**, *10*, 551. [CrossRef]
79. Grossmann, L.; McClements, D.J. Current Insights into Protein Solubility: A Review of Its Importance for Alternative Proteins. *Food Hydrocoll.* **2023**, *137*, 108416. [CrossRef]
80. Louhasakul, Y.; Cheirsilp, B.; Intasit, R.; Maneerat, S.; Saimmai, A. Enhanced Valorization of Industrial Wastes for Biodiesel Feedstocks and Biocatalyst by Lipolytic Oleaginous Yeast and Biosurfactant-Producing Bacteria. *Int. Biodeterior. Biodegrad.* **2020**, *148*, 104911. [CrossRef]
81. Amaral, P.F.F.; da Silva, J.M.; Lehocky, M.; Barros-Timmons, A.M.V.; Coelho, M.A.Z.; Marrucho, I.M.; Coutinho, J.A.P. Production and Characterization of a Bioemulsifier from *Yarrowia lipolytica*. *Process Biochem.* **2006**, *41*, 1894–1898. [CrossRef]
82. Hang, Y.; Wang, J.; Hou, Y.; Hu, S.-Q. Production of Yeast Hydrolysates by *Bacillus subtilis* Derived Enzymes and Antihypertensive Activity in Spontaneously Hypertensive Rats. *Food Biotechnol.* **2020**, *34*, 262–281. [CrossRef]
83. Guo, H.; Guo, S.; Liu, H. Antioxidant Activity and Inhibition of Ultraviolet Radiation-Induced Skin Damage of Selenium-Rich Peptide Fraction from Selenium-Rich Yeast Protein Hydrolysate. *Bioorg. Chem.* **2020**, *105*, 104431. [CrossRef] [PubMed]
84. Oliveira, A.S.; Ferreira, C.; Pereira, J.O.; Pintado, M.E.; Carvalho, A.P. Spent Brewer's Yeast (*Saccharomyces cerevisiae*) as a Potential Source of Bioactive Peptides: An Overview. *Int. J. Biol. Macromol.* **2022**, *208*, 1116–1126. [CrossRef]
85. Amorim, M.M.; Pereira, J.O.; Monteiro, K.M.; Ruiz, A.L.; Carvalho, J.E.; Pinheiro, H.; Pintado, M. Antiulcer and Antiproliferative Properties of Spent Brewer's Yeast Peptide Extracts for Incorporation into Foods. *Food Funct.* **2016**, *7*, 2331–2337. [CrossRef]
86. Branco, P.; Francisco, D.; Monteiro, M.; Almeida, M.G.; Caldeira, J.; Arneborg, N.; Prista, C.; Albergaria, H. Antimicrobial Properties and Death-Inducing Mechanisms of Saccharomycin, a Biocide Secreted by *Saccharomyces cerevisiae*. *Appl. Microbiol. Biotechnol.* **2017**, *101*, 159–171. [CrossRef]
87. Caldeira, J.; Almeida, G.; Macedo, A.L.; Silva, J.P.M.; Albergaria, H. Saccharomycin, a Biocide from *S. cerevisiae* That Kill-off Other Yeasts. *Ann. Med.* **2024**, *51*, 94–95. [CrossRef]
88. Cui, C.; Qian, Y.; Sun, W.; Zhao, H. Effects of High Solid Concentrations on the Efficacy of Enzymatic Hydrolysis of Yeast Cells and the Taste Characteristics of the Resulting Hydrolysates. *Int. J. Food Sci. Technol.* **2016**, *51*, 1298–1304. [CrossRef]
89. Niu, Y.; Gu, Y.; Zhang, J.; Sun, B.; Wu, L.; Mao, X.; Liu, Z.; Zhang, Y.; Li, K.; Zhang, Y. Characteristics of Saltiness-Enhancing Peptides Derived from Yeast Proteins and Elucidation of Their Mechanism of Action by Molecular Docking. *Food Chem.* **2024**, *449*, 139216. [CrossRef]

foods

MDPI

Article

Rapid Protein Extraction from Canola Meal Pre-Treated with Enzymatic Reactive Extrusion

Sunandita Ghosh [1], Edith Cristina González Hernández [1], Xinmei Sha [1], Jeff Chow [1], Fernanda San Martin-Gonzalez [1], Qing Jin [2] and Da Chen [1,*]

[1] Department of Food Science, Purdue University, 745 Agriculture Mall Drive, West Lafayette, IN 47907, USA
[2] School of Food and Agriculture, University of Maine, Orono, ME 04469, USA
* Correspondence: chen3370@purdue.edu

Abstract

Conventional alkaline extraction of plant proteins typically requires highly alkaline conditions (pH $\geq$ 11) and extended extraction times (~1 h). Although protease addition can lower extraction pH and improve functionality, it often requires prolonged hydrolysis. In this study, enzymatic reactive extrusion (*eREX*) using Alcalase, followed by a short duration alkaline extraction (5 min, pH 9), was evaluated as an alternative approach for producing protein-rich extracts from canola meal. The *eREX* process increased protein recovery by 48% and 42% compared with alkaline extraction conducted without and with Alcalase, respectively. The resulting powdered extracts reached a protein content of up to 49% and consisted primarily of partially hydrolyzed proteins (10–23 kDa) with increased surface hydrophobicity. Amino acid analysis showed substantial enrichment of essential amino acids, particularly histidine and sulfur-containing amino acids. Functional properties were improved, including enhanced solubility across pH 2–10, high foaming stability (88%), and increased oil-binding capacity (~5.5 g g^{-1}), while in vitro digestibility remained comparable (~85%). Techno-economic analysis indicated reductions in water use (~11%), energy consumption (~48%), and production cost (16–25%). Overall, *eREX* provides a rapid, higher-throughput, and cost-effective strategy for producing premium canola protein ingredients.

Keywords: thermal mechanical treatment; functionality; limited hydrolysis; plant proteins; amino acid composition; in vitro digestion

1. Introduction

Plant protein-based foods are, on average, two to four times more expensive than their animal-based counterparts [1]. Although improvements in formulation and processing are gradually reducing the price disparity, the gap persists. One major contributor to the high costs of plant-based foods is the plant protein ingredients themselves, which account for over 30% of total production costs [2]. Utilizing proteins derived from food and agricultural by-products has been proposed as a cost-effective strategy to reduce ingredient costs while promoting sustainability in food production [3].

Rapeseed is the second most produced oilseed after soybean with 90 million tons production in 2023–2024 [4]. Canola is a rapeseed cultivar that has lower antinutritional factors than rapeseed [5]. Canola meal, a by-product of oil extraction, contains 35–45% proteins [6,7] and is predominantly composed of the 11S globulin cruciferin and 2S albumin napin [8]. The proteins provide over 400 mg of essential amino acids per gram of protein, including tyrosine, cysteine, and methionine [8]. The latter two range from 40 to 49 mg g^{-1} protein, which position them superior to soy, pea, and wheat proteins in terms of amino acid profile.

Many approaches have been used to extract proteins from canola/rapeseed meal to produce isolate or concentrate, with alkaline–isoelectric point precipitation being widely adopted [5]. A pH value of 10–12 and mixing time of 0.5–1 h is commonly required to extract 40–70% proteins from canola/rapeseed meals [5,9]. The yield increases with the increase of pH value, but higher pH accelerates protein denaturation, uncontrolled hydrolysis, and lysinoalanine formation, resulting in diminished functionality and nutritional value [10]. Adding protease during alkaline extraction enables milder alkaline condition while raising protein recovery [11]. The resulting hydrolysates may also exhibit superior functionality such as solubility, oil absorption, emulsification, and foaming properties [10]. However, enzyme-assisted extraction requires 1–3 h hydrolysis time to achieve considerable yields [11]. It typically occurs in suspensions with low solid content ($\leq$7%), leading to high energy consumption for proteolysis and enzyme inactivation (85–100 °C). Such prolonged hydrolysis also generates small peptides imparting bitterness.

To shorten the cycle time of alkaline-based and/or enzyme-assisted plant protein extraction, mass-transfer limitations should be addressed by accelerating protein diffusion into extraction medium and minimizing the lag between extraction and separation. Twin-screw extrusion emerges as a suitable intensification platform. Rotating screws within a heated, close-tolerance barrel delivers controlled shear, mixing, and efficient heat transfer, which fragments the feed and increases the interfacial area with the surrounding medium, thereby facilitating solubilization. This strategy is well documented in lignocellulosic biorefineries under alkaline or ionic-liquid conditions [12–14]. By contrast, extrusion has rarely been used as a stand-alone unit operation for direct plant protein extraction, aside from alfalfa leaves where in-barrel mechanical expression yields protein-containing juice [15,16]. This is mainly due to the short residence time of extrusion that inhibits sufficient alkaline solubilization of proteins from raw plant matrices.

Reducing the surface hydrophobicity and lowering the molecular weight of plant proteins promotes alkaline solubilization by increasing protein–water interactions. Limited proteolysis has been found to accomplish this in plant proteins [17]. When exogenous enzymes are dosed into extruder, known as enzymatic reactive extrusion (*eREX*), enzyme–substrate contact is intensified in high-solid matrix ($\geq$20%), compressing reaction time from hours to minutes relative to stirred-tank hydrolysis [18,19]. *eREX* has been used for continuous production of protein hydrolysates with specified degrees of hydrolysis and high solubility/interfacial activity for high-protein beverages, emulsified foods, and clinical/sports nutrition [20–22]. However, applying *eREX* specifically to accelerate solubilization for plant-protein extraction, and thereby shortening extraction time and enhancing recovery yield, remains largely unexplored. By combining thermal–mechanical treatment and proteolysis, we hypothesize that *eREX* enables degrading plant proteins during the extrusion process to accelerate its diffusion under alkaline environment for rapid extraction.

In the present study, Alcalase was incorporated during extrusion of canola meal to facilitate partial protein hydrolysis, followed by a short-duration alkaline extraction. The protein recovery, physicochemical characteristics, and functional attributes, and the in vitro digestibility of the extracted canola proteins, were examined in comparison with alkaline extraction with and without Alcalase. A detailed techno-economic analysis was also conducted to quantify process efficiency and assess the feasibility of integrating this approach into larger-scale operations. The findings demonstrate enzymatic reactive extrusion could markedly shorten protein extraction time with

https://doi.org/10.3390/foods15030498

a higher plant protein recovery, which potentially reduces production cost for wide incorporation in foods.

2. Materials and Methods

2.1. Materials

Solvent-extracted canola meal with 38.5% protein, 27.6% carbohydrate, 9.1% moisture, 5.9% ash, 1.2% phenol, and 17.7% others (lignin, etc.) was ground to pass through a 60-mesh screen prior to use. Alcalase (Catalog no: 126741, from *Bacillus licheniformis*, ≥0.75 Anson units/mL), bile bovine (B3883), pancreatin from porcine pancreas (P7545), and Pefabloc SC (76307) were purchased from Sigma Aldrich (St. Louis, MO, USA). Alpha amylase and pepsin were obtained from MP Biochemicals and Mallinckrodt AR (ACS), respectively. Mini-Protean TGX Stain-Free Precast Gels 4–20% (4568095), 10× Tris/Tricine/SDS running buffer (1610744), tricine sample buffer (1610739), Precision plus pre-stained protein standards of 2–250 kDa (1610377), and Bio-safe Coomassie Blue G-250 (1610787) were purchased from Bio-Rad Laboratories (Hercules, CA, USA). All other chemicals used were from Fisher Scientific (Waltham, MA, USA). Water used was deionized.

2.2. Extrusion Treatment

The extrusion process was carried out in a Brabender TwinLab-F 20/40 extruder. The equipment has two screws of 20 mm of diameter and 795 mm screw length. The extruder barrel temperature was set to 55 °C across the first four zones based on the optimum temperature of Alcalase [23]. The fifth zone (melting zone) was set to 110 °C to inactivate Alcalase. A round strand die head (100 mm × 125 mm) without nozzle was used at the end of the barrel for maintaining the diameter uniformity in the extrudates. The feed rate of canola meal was maintained at 2 kg/h and the screed speed was set at 200 rpm to balance the mixing and residence time. Alkaline water (pH 11) with Alcalase was injected through a pump with a moisture of 50%, 60%, and 65%. The Alcalase to dry canola meal ratio was maintained at 0.1% (v/w) based on our preliminary study (Figure S1). Canola meal without Alcalase were also extruded as control (Ex-Ctrl). The residence time was measured around 2.5 min and the pressure ranged between 20–25 psi.

2.3. Protein Extraction

Canola meal extrudates, with and without Alcalase treatment, were dispersed in water to a solids content of 10% (w/v), adjusted to pH 9.0, stirred for 5 min at ambient temperature, and centrifuged (15,000× g, 30 min) to obtain the supernatant fraction containing proteins. Non-extruded canola meal at the same solids content, with and without Alcalase, were also alkaline extracted at 55 °C to approximate the thermal conditions used during extrusion to facilitate Alcalase hydrolysis. Alcalase was added at 0.1% (v/w) enzyme to dry canola meal ratio, stirred for 5 min, and then heated at 100 °C for 3 min to inactivate the enzyme prior to centrifugation. The protein content of all supernatants was quantified using the modified Lowry protein assay, and protein recovery was calculated relative to the total protein present in the original canola meal. The extrudates with the highest protein recovery were chosen for further analysis. Protein extracts obtained from alkaline extraction of non-extruded canola meal, non-extruded canola meal with Alcalase, extruded canola meal, and extruded canola meal with Alcalase were designated as Ctrl, Ctrl-Alc, Ex-Ctrl, and Ex-Alc, respectively.

2.4. Proximate Analysis and Phenol Content

The proximate composition of the freeze-dried extract powder was determined following established protocols with slight modifications. Moisture content was assessed

https://doi.org/10.3390/foods15030498

by heating the samples at 105 °C in an oven until a consistent weight was obtained. The ash content of the samples was determined in a muffle furnace at 550 °C for 20 h until a white–grey residue was obtained [24]. The total protein content was measured using the Kjeldahl method with a nitrogen-to-protein conversion factor of 5.7. The carbohydrates were extracted by a two-step hydrolyzing method as outlined by the National Renewable Energy Laboratory (NREL) [25] followed by determination using the phenol-sulfuric acid assay [24]. The lipid content was measured using the vanillin reagent method [26]. The total phenols in were determined by Folin Ciocalteu reagent phenols after extracting using 80% methanol [27]. The other components were determined by the percentage difference.

2.5. *Characterization of the Proteins in Extract*

2.5.1. Degree of Hydrolysis

The degree of hydrolysis (DH) was calculated by using the *o*-phthalaldehyde (OPA) method [17]. Briefly, 5% suspensions of the extracts were mixed with OPA reagent for 2 min before measuring the absorbance at 340 nm. The complete hydrolysis of the proteins was also conducted by heating at 120 °C for 24 h under 6 N HCl. The DH was calculated as the ratio of the sample absorbance to that of complete hydrolysis and expressed in percentage.

2.5.2. Molecular Weight Profile Determined by SDS-PAGE

The samples were dissolved in 1× Tris running buffer to obtain a final protein content of 4 mg mL^{-1}. A 100 µL aliquot of these solutions was mixed with 95 µL tricine sample buffer and 5 µL of 2-Mercaptoethanol as the reducing agent. The mixture was heated in a boiling water bath for 10 min and then centrifuged (10,000× *g*, 15 min). The supernatants (10 µL) and marker (8 µL) were carefully loaded on to the 4–20% gels and electrophoresed at 200 V for 20 min. The gel was fixed, washed with water, stained with Coomassie blue and de-stained with dilute acetic acid before imaging using GelDoc Go Imaging System (Biorad, Hercules, CA, USA).

2.5.3. Amino Acid Composition

The extracts (50 mg) were hydrolyzed with 10 mL of 6 N HCl containing 0.2 mL of 25 mM α-aminobutyric acid (AABA) as an internal standard and three drops of phenol. Samples were purged with nitrogen, sealed, and incubated at 105 °C for 24 h. After cooling, hydrolysates were diluted to 50 mL with distilled water, filtered through a 0.2 µm PTFE membrane, and freeze-dried. The dried samples were reconstituted in 1 mL of citric acid buffer (pH 2.2). Amino acid calibration standards (0–250 µmol/L) were prepared with 0.1 mM AABA as an internal standard. For derivatization, 10 µL of either standard or sample was mixed with 70 µL of AccQ•Fluor borate buffer and 20 µL of reconstituted AccQ•Fluor reagent, vortexed, and incubated at 55 °C for 10–15 min. Separation was performed on an Agilent 1200 HPLC system using an AccQ•Tag amino acid column (Nova-Pak C18, 4 µm, 150 × 3.9 mm, Waters) with a diode-array detector at 260 nm under a gradient elution of Eluent A (AccQ•Tag concentrate in water) and Eluent B (acetonitrile and 0.1% formic acid).

2.5.4. Surface Hydrophobicity

The surface hydrophobicity (H_0) was measured as described previously [28]. Samples were prepared at pH 7 to achieve various protein concentrations (0.01–0.1%, *w*/*v*), reacted with 2 mM 8-Anilino-1-naphthalenesulfonic acid (ANS) in the dark for 15 min before measuring fluorescence intensity at 390 nm excitation wavelengths of 390 nm and 480 nm and emission wavelengths. The H_0 value for each sample was determined as the slope of the linear correlation between protein concentration and fluorescence intensity.

https://doi.org/10.3390/foods15030498

2.5.4.1. ζ-Potential and Protein Solubility

The extract powders were dispersed in water to achieve a 1% (*w*/*v*) concentration. The pH of the suspensions was adjusted to 2–11 using either HCl or NaOH, stirred continuously at 25 °C for 30 min, and centrifuged at 10,000× *g* for 15 min. The ζ-potential of the above supernatants was determined using a Zetasizer (Malvern Panalytical, Worcestershire, UK) with a backscattering angle of 173° at 25 °C. The refraction index of protein used was 1.45. The protein content in the supernatants was measured by using modified Lowry Assay. The protein solubility was calculated as the ratio of the amount of the protein in the supernatant to the total protein content in the sample and expressed as percentage.

2.6. Functional Properties

2.6.1. Foaming Capacity and Stability

The extracts (0.5 g) were mixed with 20 mL water using a homogenizer at 8000 rpm for 30 s. They were further homogenized at 10,000 rpm for 3 min in an ice bath. The volume of the foam formed was recorded for the foaming capacity (FC). To measure the foaming stability (FS) the volume of the foam remaining after 90 min was also recorded. The foaming capacity and stability of the samples were calculated using Equations (1) and (2) [29].

$$\text{Foaming capacity (FC) (\%)} = \frac{Initial\ volume\ (\text{mL})}{Volume\ of\ foam\ (\text{mL})} \times 100 \quad (1)$$

$$\text{Foaming stability (FS) (\%)} = \frac{Initial\ volume\ (\text{mL})}{Volume\ of\ foam\ after\ 90\ \text{min}\ (\text{mL})} \times 100 \quad (2)$$

2.6.2. Water-Holding and Oil-Binding Capacity

The extracts (0.5 g) were mixed with water (5 g) and left to stand for 10 min and then vortexed for 10 sec [30]. This was repeated six times, then the extracts were centrifuged at 1000× *g* for 15 min. The pellet obtained was used to determine the water-holding capacity (WHC) using Equation (3). This process was repeated using oil instead of water to determine the oil-binding capacity (OBC).

$$\text{WHC or OBC (\%)} = \frac{Weight\ of\ pellet(\text{g}) - Weight\ of\ sample\ (\text{g})}{Weight\ of\ sample\ (\text{g})} \times 100 \quad (3)$$

2.6.3. Emulsification Properties

The emulsification capacity (EC) was assessed based on the initial droplet size formed upon emulsification, while the ES was determined by monitoring changes in droplet size during 7 days of storage. Emulsions were made as per the procedure described in [29]. Briefly, extract suspensions of 1% (*w*/*v*) were prepared in pH 7 phosphate buffer (10 mM) containing 0.02% (*w*/*v*) sodium azide to prevent microbial growth. Pre-emulsions were prepared by homogenizing 3 mL soybean oil with 30 mL of protein suspension at 12,000 rpm for 1 min. Emulsification was carried out by further homogenization using a Nano DeBee high pressure homogenizer (BEE International, South Easton, MA, USA) at 5500 psi for 10 passes. The emulsions were collected by placing them in an ice bath and stored at 4 °C for further analysis. The emulsion mean-diameter (Z-Average) was measured on day 0 (fresh), 1, 3, and 7 using a Zetasizer (Malvern Panalytical, Worcestershire, UK) with a backscattering angle of 173° at 25 °C. The refractive index for protein used was 1.45.

https://doi.org/10.3390/foods15030498

2.6.4. Gelation Dynamics

DHR-3 hybrid rheometer (TA Instruments, New Castle, DE, USA) was used to study the gelation dynamics at 15% (w/v) extract concentration. A 40 mm parallel plate geometry with a 1 mm gap was employed, and the geometry boundary was sealed with liquid paraffin to minimize evaporation. A temperature ramp from 25 °C to 90 °C at 5 °C/min was used, held at 90 °C for 30 min, and cooled to 25 °C at the same rate. The storage modulus (G') and loss modulus (G'') were recorded throughout the process under 0.5% strain (within linear viscoelastic region) and 1 Hz frequency. At least triplicates were run for each sample.

2.7. *In Vitro Digestion*

In vitro digestion of the proteins was carried out following the INFOGEST protocol [31]. The enzymes, bile, and $CaCl_2$ were freshly prepared before the experiment. The final enzyme activity of salivary amylase, pepsin, lipase, and pancreatin were 75 U mg^{-1}, 2000 U mg^{-1}, 60 U mg^{-1}, and 100 U mg^{-1}. The final concentration of bile used was 10 mM. Suspensions of extracts at 10% (w/v) were prepared and left overnight at 4 °C. To 2.5 mL of the sample, 2.5 mL of simulated salivary fluid was added and slowly stirred at 37 °C for 2 min. Simulated gastric fluid (5 mL) was added to this mixture and again slowly stirred at 37 °C for 2 h. The pH of the mixture was maintained at 3. Next, 10 mL simulated salivary fluid was added, the pH was maintained at 7 and allowed to sit for 2 h at 37 °C under slow stirring. To inactivate the enzymes, Pefabloc at 5 mM final concentration was added at the end of the digestion. The total soluble proteins were obtained from the supernatant after centrifugation (10,000× g, 15 min) of the digesta. The apparent digestibility of the proteins was determined by Equation (4).

$$\text{Apparent digestibility } (\%) = \frac{Total\ soluble\ proteins\ after\ digestion}{Total\ proteins\ before\ digestion} \times 100 \quad (4)$$

2.8. *Techno-Economic Analysis*

The TEA model was built using SuperPro Designer (Version 14, Intelligen, Inc., Scotch Plains, NJ, USA), in which mass and energy flows were tracked. The plant was designed to process 50,000 MT of canola meal per year, corresponding to a middle-sized plant. The production plant was scheduled to operate 330 days per year and 24 h per day. Equipment costs were obtained from previous models and the build-in cost in SuperPro Designer software (Version 14). Equipment costs were scaled to reflect differences in processing capacity using the customary exponential scaling relationship (Equation (5)). Costs were then updated to 2023 dollars using the Chemical Engineering Plant Cost Index (CEPCI), as shown in Equation (6):

$$Cost_{new} = Cost_{base} \times (Size_{new}/Size_{base})^{0.6} \quad (5)$$

$$Cost_{2023} = Cost_{base} \times (CEPCI_{2023}/CEPCI_{base}) \quad (6)$$

The total capital investment (TCI) was estimated as the sum of the fixed capital investment (FCI), working capital (assumed as 10% of FCI), and land cost (5% of FCI). The FCI comprises both direct expenses, such as equipment purchase, installation, instrumentation, piping, building structures, electrical systems, utilities, and site preparation, and indirect expenses, including construction overhead, engineering and supervision, contractor fees, legal costs, and contingencies. These categories were derived as standard percentage allocations relative to equipment cost. Table S1 summarizes the breakdown of variable

https://doi.org/10.3390/foods15030498

and fixed operating expenses. Variable costs consist of chemicals, enzymes, utilities, and waste management, with values sourced from earlier studies (Table S1). Fixed operating expenses, such as labor, associated overhead, maintenance, insurance, and depreciation were estimated. After determining the total operating cost, the unit cost of protein production was calculated by dividing the annual operating cost by the yearly amount of recoverable protein product.

2.9. Statistical Analysis

The results were analysed using average and standard deviation of the measurements. All the experiments were carried out in either duplicates or triplicates. Statistical analysis was conducted by using analysis of variance (ANOVA) with the post hoc Tukey's HSD significant difference test using OriginPro 2023b (Northampton, MA, USA). Significance was established at $p < 0.05$.

3. Results and Discussions

3.1. Extraction, Protein Yield, and Proximate Composition

All the extrudates showed a uniform dark-brown color and cylindrical shape (Figure 1A). At 50% moisture, both extrudates ('Without and With Alcalase') exhibited surface fissures (Figure S2) due to limited plasticization. The shape intactness decreased with an increase in moisture. Following alkaline extraction, the soluble protein content in the extracts was higher for the extrudates treated with Alcalase, regardless of the initial moisture content (Figure 1B), with the highest value at 65% moisture. Increasing the moisture content could reduce canola meal matrix viscosity and enhance enzyme diffusion. It also promoted swelling of the dry meal particles, creating additional pores, and increasing the enzyme–protein contact surface. The elevated moisture levels could also accelerate protein hydration and facilitate its diffusion into water phase, thus facilitating protein extraction. Therefore, extrudate moisture at 65% was selected for subsequent analyses.

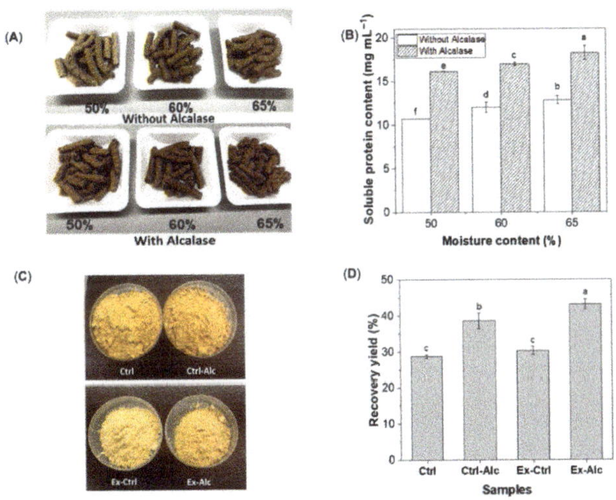

Figure 1. (**A**) Photographs of canola meal extrudates at different moisture content without and with Alcalase; (**B**) The soluble protein content of the extrudates at different moisture content; (**C**) Photographs of freeze-dried extracts; (**D**) Protein recovery yield of extracts. Ctrl, alkaline extraction without Alcalase; Ctrl-Alc, alkaline extraction with Alcalase; Ex-Ctrl, Alkaline extraction of extruded samples without Alcalase; Ex-Alc, alkaline extraction of extruded samples with Alcalase. Different letters on top of the bars indicate significant differences among the data ($p < 0.05$).

https://doi.org/10.3390/foods15030498

The extracts obtained from both non-extruded and extruded canola meal were freeze-dried. The extruded samples had a lighter brown color than their non-extruded counterparts (Figure 1C), possibly due to the partial degradation of pigments by extrusion [32]. Enzymatic hydrolysis (Ctrl-Alc) improved the protein recovery yield by approximately 35% compared to the untreated control (Ctrl) (Figure 1D). Whereas, extrusion alone (Ex-Ctrl) had no effect on the protein recovery yield compared to the Ctrl, the yield increased substantially, by ~50%, with Alcalase included (Ex-Alc), representing the highest protein recovery yield among all the extracts.

Proximate analysis of the extracts showed 49% proteins in Ex-Alc sample, which is significantly higher than those from Ctrl, Ctrl-Alc, and Ex-Ctrl samples (Table 1). The syneresis of proteolysis and extrusion-induced porosity and meal particle degradation (Figure S3) would enhance the alkaline solubility and diffusion [33], resulting in a higher protein purity in the extract. Carbohydrate levels remained relatively constant (27–28%) across all extracts. The carbohydrates are most likely to be pectins and hemicellulose, as they can be extracted by an alkaline medium. Increase in ash content was detected following extrusion, which may be attributed to the breakdown of matrix components that release minerals in the extractable phase [34]. The lipid content significantly dropped upon extrusion (both for Ex-Ctrl and Extr, With Alc). The "others" fraction ranged from 6–23% among the samples, which probably represents lignin. The value was the lowest in *eREX* extract due to the relatively increase of other components.

Table 1. Proximate analysis (%) of freeze-dried canola meal extracts. Ctrl, alkaline extraction without Alcalase; Ctrl-Alc, alkaline extraction with Alcalase; Ex-Ctrl, alkaline extraction of extruded samples without Alcalase; Ex-Alc, alkaline extraction of extruded samples with Alcalase. 'Others' is primarily composed of acid-insoluble lignin. Different superscript letters indicate significant differences among the data ($p < 0.05$).

Samples	Moisture	Ash	Protein	Carbohydrate	Lipid	Others	Phenol
Ctrl	4.6 ± 0.0^{b}	10.8 ± 0.2^{b}	31.4 ± 0.1^{c}	27.8 ± 0.2^{a}	0.13 ± 0.01^{b}	22.5 ± 0.3^{a}	2.6 ± 0.1^{b}
Ctrl-Alc	5.3 ± 0.2^{a}	10.7 ± 0.2^{b}	37.9 ± 0.1^{b}	28.2 ± 0.6^{a}	0.17 ± 0.00^{a}	14.9 ± 0.7^{b}	2.8 ± 0.1^{ab}
Ex-Ctrl	2.7 ± 0.1^{d}	12.5 ± 0.2^{a}	32.1 ± 0.8^{c}	27.5 ± 0.0^{a}	0.04 ± 0.00^{c}	22.7 ± 0.9^{a}	2.4 ± 0.0^{c}
Ex-Alc	3.6 ± 0.2^{c}	11.8 ± 0.2^{a}	49.2 ± 0.7^{a}	26.6 ± 1.1^{b}	0.03 ± 0.00^{c}	6.0 ± 1.3^{c}	2.8 ± 0.0^{a}

In conventional alkaline extraction, approximately 50–80% of the alkaline-soluble proteins precipitate during isoelectric precipitation (pH 4.5–5.5), while the remaining acid-soluble proteins are typically discarded unless recovered through membrane filtration. Although acid precipitation increases protein purity, it reduces overall recovery and elevates processing costs. In contrast, the *eREX* process omits the acid-precipitation step, producing a protein-rich extract that retains nearly all the alkaline soluble proteins and other valuable non-protein components such as carbohydrate, although by compromising protein purity. The developed protein extract is thus more applicable as protein concentrate rather than isolate. As shown in Figure S4, a 5 min extraction following *eREX* pretreatment achieved protein yields comparable to a 1 h extraction in conventional Alcalase-assisted processing, demonstrating the efficiency of the *eREX* process on facilitating protein extraction.

3.2. Characterization of the Proteins in Extract

3.2.1. Degree of Hydrolysis and Molecular Weight

The proteins in Ctrl sample showed a DH of 15% (Figure 2A). A similar DH value (15%) was observed for Ex-Ctrl sample, suggesting that extrusion does not trigger peptide bond cleavage. Alcalase treatment (Ctrl-Alc) increased the DH to 23%, and *eREX* pretreatment (Ex-Alc) achieved a DH of 18%. A lower DH observed in Ex-Alc sample is probably due

 https://doi.org/10.3390/foods15030498

to a higher solid content that limited the diffusion of Alcalase into the canola meal for a lower hydrolysing efficiency. The OPA reagent used for DH measurement reacts with free amino groups; therefore, the measured absorbance reflects the abundance of these groups rather than directly quantifying the number of peptide bonds cleaved. The relatively high initial DH obtained using the OPA method likely reflects its sensitivity to trace levels of pre-existing low-molecular-weight peptides and free amino acids produced during the industrial oil extraction process [7], in addition to the inherent reactivity of lysine ε-amino groups. Nevertheless, the relative higher values of DH in Ex-Alc and Ctrl-Alc suggest an effective enzymatic hydrolysis is 3% and 8%, respectively.

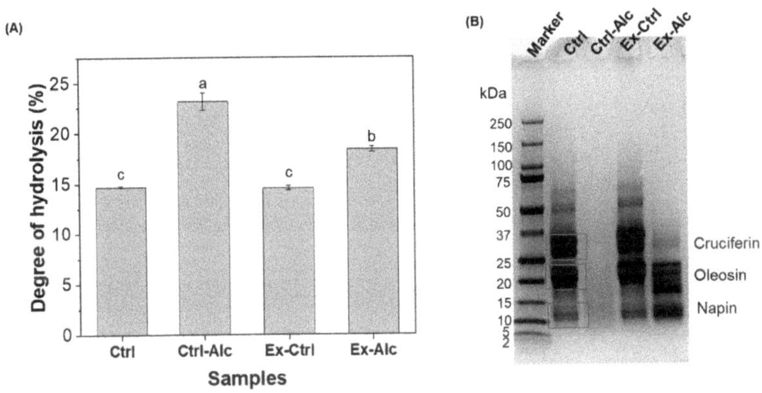

Figure 2. (**A**) Degree of Hydrolysis; (**B**) SDS profile of protein in the canola meal extracts. Ctrl, alkaline extraction without Alcalase; Ctrl-Alc, alkaline extraction with Alcalase; Ex-Ctrl, Alkaline extraction of extruded samples without Alcalase; Ex-Alc, alkaline extraction of extruded samples with Alcalase. Different letters on top of the bars indicate significant differences among the data ($p < 0.05$). The boxes highlight the corresponding proteins of the bands.

The SDS-PAGE profile of the proteins in extract showed Ctrl sample had protein bands within 10–75 kDa (Figure 2B) with prominent ones located at around 30 kDa and 20 kDa corresponding to the α- and β-chains of cruciferin [35], and oleosins [8] respectively. Napin subunits were also detected at approximately 12 kDa [35]. Ex-Ctrl showed a profile comparable to the Ctrl sample, which agrees well with their similar DH values. Inclusion of Alcalase (Ex-Alc) generated peptide fragments primarily within the 10–23 kDa range indicate partial hydrolysis of the proteins. The bands were not visible with the appearance of a smear region in Ctrl-Alc sample, corresponding to a higher DH value.

3.2.2. Amino Acid Composition

Amino acid profiles were analyzed to assess the nutritional attributes of proteins obtained from different extraction approaches (Table 2). The total amino acid content was lower than the protein content measured by Kjeldahl, which can be explained by methodological differences. The Kjeldahl assay quantifies total nitrogen and may overestimate protein content by up to 6% [36] due to the inclusion of non-protein nitrogen compounds, whereas amino acid analysis can underestimate total amino acids because tryptophan is destroyed and certain residues, such as lysine, cysteine, and methionine, undergo partial degradation during acid hydrolysis [37]. The total amino acid concentration increased from 241 mg g^{-1} extract in the Ctrl sample to 291 mg g^{-1} in Ctrl-Alc, and 386 mg g^{-1} in Ex-Alc. The substantial increase in total amino acids following *eREX* aligns with higher protein purity (Table 1).

https://doi.org/10.3390/foods15030498

Table 2. Amino acid composition of extracts from canola meal. Ctrl, alkaline extraction without Alcalase; Ctrl-Alc, alkaline extraction with Alcalase; Ex-Ctrl, Alkaline extraction of extruded samples without Alcalase; Ex-Alc, alkaline extraction of extruded samples with Alcalase. AA (amino acid); Asx (asparagine + aspartic acid) and Glx (glutamine + glutamic acid). The data were expressed as mean ± SD. Different superscript letters within the same row indicate significant differences among the data ($p < 0.05$).

Amino Acid (mg g^{-1} Extract)	Ctrl	Ctrl-Alc	Ex-Ctrl	Ex-Alc
Essential AA				
His	7.9 ± 0.4^{cd}	9.3 ± 0.9^{c}	17.8 ± 2.7^{b}	31.7 ± 2.0^{a}
Ile	7.6 ± 0.6^{b}	7.2 ± 0.4^{b}	6.4 ± 0.9^{b}	10.7 ± 0.2^{a}
Leu	14.0 ± 1.3^{b}	15.1 ± 0.9^{b}	12.5 ± 1.1^{b}	20.2 ± 0.3^{a}
Lys	3.5 ± 0.3^{c}	4.5 ± 0.3^{b}	2.5 ± 0.2^{d}	5.6 ± 0.4^{a}
Met	5.2 ± 0.4^{c}	6.5 ± 0.4^{b}	4.4 ± 0.4^{c}	7.8 ± 0.2^{a}
Phe	9.1 ± 0.8^{a}	8.0 ± 0.5^{a}	8.2 ± 0.6^{a}	11.4 ± 0.2^{a}
Val	6.0 ± 0.5^{c}	9.4 ± 0.6^{b}	5.8 ± 0.5^{c}	12.1 ± 0.2^{a}
Thr	8.4 ± 0.6^{c}	11.6 ± 0.6^{b}	7.4 ± 0.4^{c}	13.8 ± 0.1^{a}
Total essential AA	61.2 ± 1.1^{c}	71.6 ± 1.0^{b}	65.1 ± 1.8^{c}	113.3 ± 1.2^{a}
Non-essential AA				
Ala	8.6 ± 0.6^{c}	11.2 ± 0.7^{b}	7.9 ± 0.4^{c}	14.2 ± 0.6^{a}
Arg	9.5 ± 1.2^{a}	10.8 ± 0.6^{a}	7.5 ± 0.7^{a}	12.9 ± 0.5^{a}
Asx	56.0 ± 3.8^{bc}	61.5 ± 3.4^{b}	50.4 ± 1.3^{c}	78.8 ± 1.3^{a}
Cys	4.0 ± 0.3^{b}	4.7 ± 0.3^{ab}	3.8 ± 0.2^{b}	6.3 ± 0.2^{a}
Glx	12.2 ± 0.9^{b}	12.1 ± 0.7^{b}	11.1 ± 0.9^{b}	17.9 ± 0.4^{a}
Gly	7.6 ± 0.6^{bc}	9.0 ± 0.4^{b}	6.6 ± 0.4^{c}	11.2 ± 0.2^{a}
Pro	12.6 ± 0.7^{c}	16.9 ± 0.8^{b}	10.8 ± 0.6^{c}	20.2 ± 0.7^{a}
Ser	56.7 ± 5.4^{b}	78.5 ± 7.5^{a}	52.3 ± 5.4^{b}	91.22 ± 10.9^{a}
Tyr	12.7 ± 1.1^{b}	14.9 ± 0.9^{b}	11.3 ± 0.9^{b}	20.1 ± 0.3^{a}
Total Non-essential AA	179.9 ± 4.0^{c}	219.5 ± 4.9^{b}	161.6 ± 3.4^{d}	272.8 ± 6.4^{a}
Total AA	241.1 ± 4.2^{c}	291.1 ± 4.9^{b}	226.6 ± 3.8^{c}	386.0 ± 6.5^{a}

Among individual amino acids, the most notable improvement was observed for histidine, which increased from 7.9 mg g^{-1} in Ctrl to 32 mg g^{-1} in Ex-Alc. Lysine content also increased from 3.5 mg g^{-1} to 5.6 mg g^{-1} exceeding that achieved by Alcalase treatment alone (Ctrl-Alc). This is consistent with previous reports indicating that partial proteolysis prior to or during thermal processing can reduce the extent of lysine loss by disrupting protein matrices and limiting Maillard-type condensation between lysine and reducing sugars [38]. The total sulfur-containing amino acids (cysteine + methionine) also increased from 9.2 mg g^{-1} in the Ctrl sample to 11.1 mg g^{-1} in Ctrl-Alc and 14.2 mg g^{-1} in Ex-Alc, which aligns with studies showing that enzymatic pre-treatment can enhance the release of sulfur-bearing residues that are otherwise partially inaccessible within protein structures [39]. The absence of reductions in any individual amino acid content indicates that *eREX* primarily enhanced amino acid availability rather than promoting thermal degradation or oxidation, consistent with previous study that high-moisture extrusion preserved heat-labile amino acids by reducing residence time under mild thermal conditions [40].

3.2.3. Surface Hydrophobicity

The surface hydrophobicity (H_0) of the extracts was determined to evaluate structural modifications induced by different extracting approaches. Extrusion (Ex-Ctrl) resulted in a significant increase in H_0 compared to the Ctrl counterpart (Figure 3A), consistent with previous observations on soy proteins [41]. This increase can be attributed to the exposure of hydrophobic residues that were previously buried within the protein interior, due to

https://doi.org/10.3390/foods15030498

molecular unfolding during extrusion. Alcalase hydrolysis (Ctrl-Alc) decreased the H_0 compared to the Ctrl sample. Similar observations have been reported in pea hydrolysates obtained using Alcalase [17]. The H_0 of Ex-Alc samples were higher than the Ctrl but lower than Ex-Ctrl. Alcalase is known to release peptides enriched in hydrophobic amino acids [42]. These peptides may undergo conformational rearrangement or aggregation during *eREX*, leading to the re-burying of the hydrophobic side chains and a relative increase in the surface hydrophilicity.

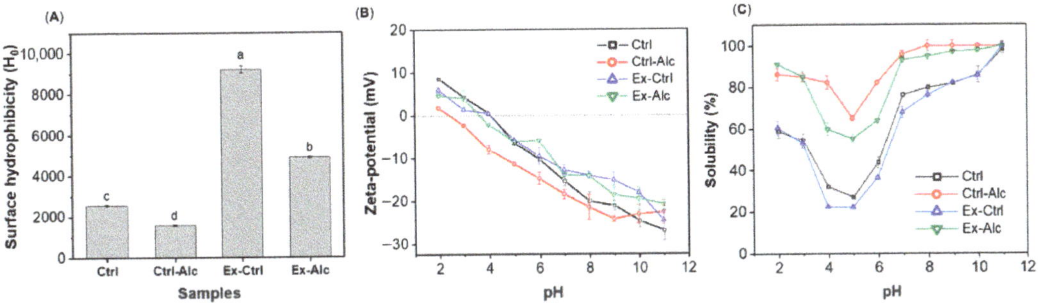

Figure 3. (**A**) Surface hydrophobicity. Different letters (on top of the bars) indicate significant differences among the data ($p < 0.05$); (**B**) Zeta-potential; (**C**) Solubility of protein extracts. Ctrl, alkaline extraction without Alcalase; Alc, alkaline extraction with Alcalase; Ex-Ctrl, Alkaline extraction of extruded samples without Alcalase; Ex-Alc, alkaline extraction of extruded samples with Alcalase.

3.2.3.1. ζ-Potential and Protein Solubility

Besides surface hydrophobicity, surface charge is another indicator that affects protein physiochemical properties and functionality. The ζ-potential values decreased with an increase in pH for all samples (Figure 3B). The Ctrl sample crossed zero around pH 4, which is consistent with the reported isoelectric point (pI) for rapeseed/canola proteins [43]. Alcalase hydrolysis (Ctrl-Alc) increased the ζ-potential values across most of the pH range (≤pH 10), consistent with proteolysis exposing ionizable residues and yielding peptides with more surface charge [44]. The Ex-Alc sample showed comparable ζ-potential values to that of Ex-Ctrl. For protein solubility, all samples showed a U-shape, with lower values between pH 4 and 6, and higher values in more acid or alkaline environments. This is within expectation, as plant proteins usually exhibit the minimal solubility at pH close to the isoelectric point (pH 4.5–5). When the pH is far away from isoelectric point, it becomes more protonated or deprotonated, exhibiting higher affinity to water. It must be noted that the ζ-potential did not show the same U-shape as the protein solubility. Since the protein extract contains other non-protein ingredients, such as cell wall polysaccharides, which contributed to the overall zeta potential values. Ex-Ctrl sample had the lowest protein solubility among the samples, consistent with its higher surface hydrophobicity (Figure 3A). The Ctrl-Alc and Ex-Alc samples significantly improved their solubility across the measured pH range. The improvement can be attributed to both reduction in molecular weight and increased exposure of ionizable groups after proteolysis [44].

3.3. *Functional Properties of Extracts*

3.3.1. Foaming and Emulsification Properties

The Ctrl sample showed a foaming capacity (FC) of 16%, which aligns with a previous reported value of 18% for rapeseed proteins (18%) [45]. Alcalase treatment significantly improved the FC to 75–80% in both the non-extruded (Ctrl-Alc) and extruded (Ex-Alc) extracts. Similar improvements were observed in Alcalase-hydrolyzed rapeseed proteins [45–47]. The FS was the highest for the Ex-Alc (88%), while the others exhibited significantly lower

https://doi.org/10.3390/foods15030498

stability (~45–55%) (Figure 4B). This enhanced FC and FS suggests that Alcalase inclusion produced peptides that rapidly adsorb at the air–water interface and reduce the surface tension to sustain the foam [47].

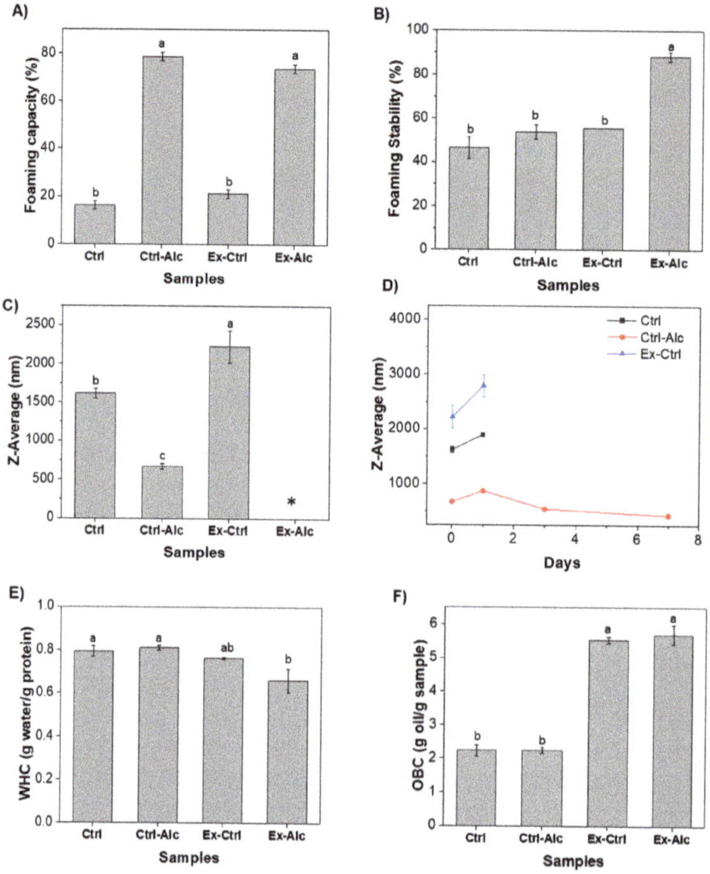

Figure 4. (**A**) Foaming capacity; (**B**) foaming stability; (**C**) emulsification capacity; (**D**) emulsification stability; (**E**) water-holding capacity (WHC); (**F**) oil-binding capacity (OBC) of canola meal extracts. Ctrl, alkaline extraction without Alcalase; Ctrl-Alc, alkaline extraction with Alcalase; Ex-Ctrl, alkaline extraction of extruded samples without Alcalase; Ex-Alc, alkaline extraction of extruded samples with Alcalase. Different letters on top of the bars indicate significant differences among the data ($p < 0.05$). (*) in (**C**) denotes droplet size was not obtained due to visible phase separation. In (**D**), by day 3 (for 'Ctrl' and 'Ex-Ctrl'), no size data could be obtained due to visible phase separation.

Emulsions produced with Ctrl sample exhibited a Z-average of ~1600 nm, comparable to those stabilized with >0.5% (*w*/*w*) canola proteins (~1500 nm) [48]. Ctrl-Alc had significantly higher EC, shown as a smaller droplet size (~700 nm). Extrusion (Ex-Ctrl) diminished the EC, especially with the inclusion of Alcalase (Ex-Alc), for which the droplet size was not measurable due to the visual phase separation (Figures S5 and S6). Compared with intact proteins, medium-sized peptides generally exhibit superior emulsifying performance because they diffuse rapidly to the interface while retaining sufficient chain length and flexibility to form cohesive, viscoelastic oil–water interfacial films [49]. However, when these peptides associate into aggregates, as may occur during *eREX*, indicated by increased surface hydrophobicity, the resulting structures possess

https://doi.org/10.3390/foods15030498

reduced mobility and limited ability to reorganize at the interface. Such aggregates are less effective in forming stable interfacial layers, leading to diminished EC [50]. The prepared emulsions were stored for 7 days to assess stability (ES). By day 1, all samples exhibited an increase in Z-average (Figure 4D), indicating the onset of flocculation [51]. By days 3 and 7, visible phase separation characterized by a serum and a cream layer was observed for Ctrl and Ex-Ctrl emulsions (Figure S4a,b). This confirmed progressive ES loss, consistent with destabilization initiated by flocculation, followed by creaming. In contrast, the Ctrl-Alc based emulsion showed a decrease in droplet size on day 3, indicating that the interfacial layer reorganized and re-equilibrated, thereby improving droplet stabilization [52].

3.3.2. Water-Holding and Oil-Binding Capacity

The WHC of the Ctrl sample was ~0.8 g water g^{-1} extract (Figure 4E), which is lower than the values reported for rapeseed and canola proteins (1.6–3.9 g g^{-1}) [53,54]. The relatively smaller value here may be attributed to the presence of non-protein constituents (such as phenolics, and minerals) that have a low capacity to retain water. The WHC was not affected by enzymatic treatment (Ctrl-Alc) or extrusion (Ex-Ctrl) treatment. However, combination of the two (Ex-Alc) slightly decreased the WHC (~0.65 g g^{-1}) (Figure 4E). This reduction may be attributed to peptide aggregation during *eREX* that bury polar amino acid residues and reduce their water accessibility [54].

For the oil-binding capacity (OBC), the Ctrl sample exhibited a value of ~2.2 g oil g^{-1} extract, which is comparable to previously reported values for rapeseed and canola proteins (2.3–2.8 g g^{-1}) [53,54]. Extrusion pretreatment led to a significant improvement in OBC for both 'Ex-Ctrl' and 'Ex-Alc' samples. This observation is consistent with findings for soybean [55] and hempseed [56] proteins, where extrusion increased the OBC by exposing hydrophobic groups that facilitate interactions with oil.

3.3.3. Gelation Properties

Differences in structural characteristics and molecular size between untreated and treated canola proteins are expected to affect their gelling capacity during heat-induced gel formation. Changes in storage (G') and loss modulus (G'') were thus recorded during the gelation process (Figure 5A–D). For all samples, G' values were slightly higher than G'' prior to heating, indicating a weak gel nature of the extract suspension due to high solid content [28]. During the heating stage (25 °C to 90 °C), a modest decline in G'' and relatively preserved G' was found. When the temperature was held at 90 °C, both G' and G'' increased, except in the Ctrl-Alc sample. The Ctrl sample increased from 0.4 to 16 Pa, whereas only 2.5–4.5 Pa increment was observed in extruded samples (Ex-Ctrl, Ex-Alc). The magnitude of increment is mainly controlled by molecular interactions. Extrusion could denature proteins and form smaller aggregates which are usually less reactive than individual proteins after secondary heating. Upon cooling, G' showed a greater increase than during the heating stage in Ctrl, Ex-Ctrl, and Ex-Alc samples. Plant protein-based hydrogel is mainly dominated by non-covalent interactions [57]. At lower temperatures, more hydrogen bonding and electrostatic interactions are formed, thereby exhibiting higher moduli. Compared to Ctrl and Ex-Ctrl samples, the Ex-Alc sample had lower molecular weight, which may limit the intermolecular interactions and formation of continuous gel network, corresponding to a weaker gel. Based on the rheological study, *eREX* pre-treatment had an inferior effect on the gelling capacity of the protein extracts. It is noteworthy that the extracts studied here contained moderate amounts of carbohydrates, which may also be involved in the gelation process. As carbohydrate levels were compara-

https://doi.org/10.3390/foods15030498

ble across all the extracts, their effects are likely similar and have limited contribution to the observed differences.

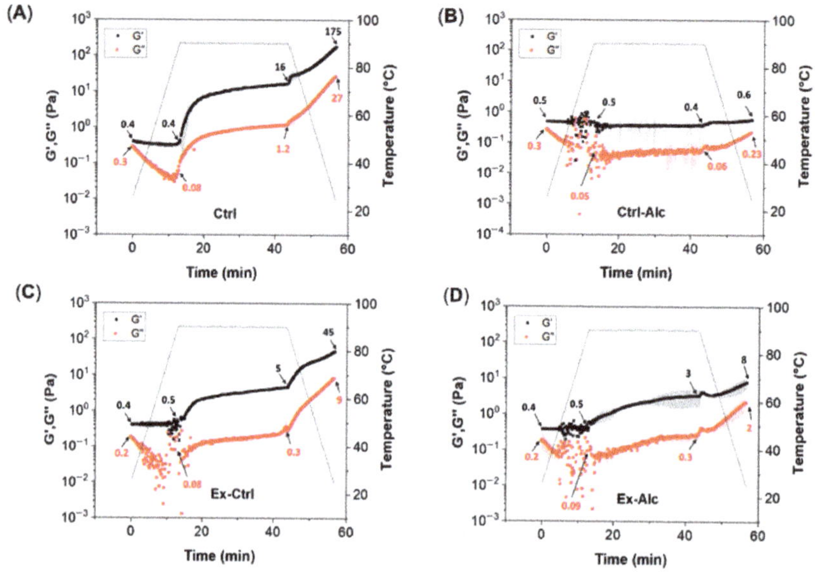

Figure 5. Gelation dynamics of canola meal extracts from different treatments. (**A**) Ctrl, alkaline extraction without Alcalase; (**B**) Ctrl-Alc, alkaline extraction with Alcalase; (**C**); Ex-Ctrl, alkaline extraction of extruded samples without Alcalase; (**D**) Ex-Alc, alkaline extraction of extruded samples with Alcalase. The number in the graphs indicates the G' and G'' values at different heating and cooling stages.

3.4. In Vitro Digestion

The apparent protein digestibility of Ctrl sample was 85%. No significant differences were observed upon Alcalase treatment (Ex-Ctrl), extrusion (Ex-Ctrl), or combination of both. To further assess whether the extracts have undergone distinct degradation, the degree of hydrolysis (DH) of the digesta was measured simultaneously. The DH of the Ctrl sample was 27% (Figure 6B) and higher than the reported value of 19% for rapeseed meal [58]. The DH of the Ex-Ctrl sample showed no significant difference to that of Ctrl sample (Figure 6A), which agrees well with the digestibility values. The DH was significantly higher for both Ctrl-Alc and Ex-Alc, with similar values between the two. This enhancement can be attributed to the greater exposure and accessibility of cleavage sites to digestive enzymes resulting from Alcalase pre-treatment. However, this increase in DH did not translate into higher apparent digestibility (Figure 6A), a trend also observed for chickpea proteins subjected to enzymatic pretreatment [59]. DH measures the extent of peptide bond cleavage, while digestibility measures the distribution of peptides between soluble and insoluble phases, despite smaller peptides usually corresponding to higher solubility. However, when DH reaches high values, the produced peptides could aggregate and precipitate, becoming insoluble. The balance between the two determines the final solubility (digestibility).

https://doi.org/10.3390/foods15030498

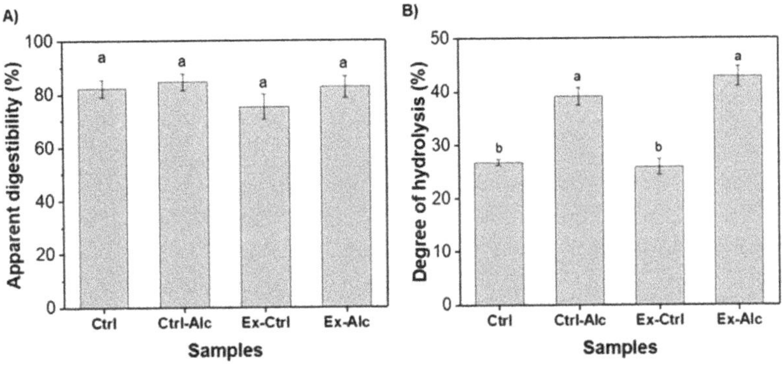

Figure 6. (**A**) Apparent digestibility and (**B**) degree of hydrolysis of in vitro digested extracts. Ctrl, alkaline extraction without Alcalase; Ctrl-Alc, alkaline extraction with Alcalase; Ex-Ctrl, Alkaline extraction of extruded samples without Alcalase; Ex-Alc, alkaline extraction of extruded samples with Alcalase. Different letters indicate significant differences among the data ($p < 0.05$).

3.5. Techno-Economic Analysis

To assess economic feasibility of large-scale production of protein concentrate using *eREX* method, a techno-economic analysis (TEA) of the process was conducted in comparison with alkaline extraction and enzyme-assisted alkaline extraction. Since extrusion-only (Ex-Ctrl) showed limited enhancement of protein recovery, it was excluded from the TEA. The total capital investment (TCI) for protein production from canola meal is summarized in Table 3. Ex-Alc shows the highest capital demand at approximately $16.4 million, largely driven by its higher equipment cost and correspondingly larger direct and indirect cost multipliers. Alkaline extraction (Ctrl) requires a moderate investment ($13.0 million), whereas Ex-Alc represents the lowest-capital option at about $10.9 million, reflecting its smaller equipment footprint and simplified processing configuration. Table 4 compares the annual operating costs and protein unit prices for alkaline extraction (Ctrl), enzymatic extraction (Ctrl-Alc), and reactive extrusion processing (Ex-Alc) plants, revealing substantial differences in economic performance among the three technologies. Ex-Alc demonstrates the lowest total operating cost ($33.2 million/year) and the lowest protein unit cost ($3.6/kg), driven primarily by reduced water consumption (due to high solid loading), lower chemical usage, and foamlower steam demand compared to the other processes. In contrast, Ctrl-Alc incurs the highest operating cost ($44.5 million/year), largely due to the high steam requirement (e.g., enzyme inactivation), reflected in utilities, contributing 44.9% of its total operating cost. Alkaline extraction shows a moderate cost profile ($39.6 million/year), but its heavy reliance on steam, large volumes of waste disposal, together with the lower protein yield keep its protein cost elevated at $10.6/kg. The findings from TEA demonstrated the economic advantage of utilizing *eREX* on canola meal protein extraction and potentially reduced the production cost.

Table 3. Total capital investment ($) for alkaline extraction (Ctrl), enzymatic extraction (Ctrl-Alc), and enzymatic reactive extrusion (Ex-Alc) processing plants.

Item	Ctrl ($)	Enzyme ($) Extraction	Extrusion ($) Extraction
Purchased equipment cost (E)	2,598,747	3,291,207	2,187,851
Purchased-equipment installation (39% E)	1,013,511	1,283,571	853,262
Instrument and controls (26% E)	675,674	855,714	568,841
Piping (31% E)	805,611	1,020,274	678,234

https://doi.org/10.3390/foods15030498

Table 3. *Cont.*

Item	Ctrl ($)	Enzyme ($) Extraction	Extrusion ($) Extraction
Electrical systems (10% E)	259,875	329,121	218,785
Buildings (29% E)	753,637	954,450	634,477
Yard improvements (12% E)	311,850	394,945	262,542
Service facilities (55% E)	1,429,311	1,810,164	1,203,318
Total direct plant costs (302% E)	7,848,215	9,939,445	6,607,310
Engineering and supervision (33% E)	857,586	1,086,098	721,991
Construction expenses (39% E)	1,013,511	1,283,571	853,262
Legal expenses (4% E)	103,950	131,648	87,514
Contractor's fee (17% E)	441,787	559,505	371,935
Contingency (35% E)	909,561	1,151,922	765,748
Total indirect costs (128% E)	3,326,396	4,212,745	2,800,449
Fixed capital investment (FCI, 430% E)	11,174,611	14,152,190	9,407,760
Working capital (10% of FCI)	1,240,382	1,570,893	1,044,261
Land (5% of FCI)	558,731	707,610	470,388
Total capital investment	12,973,723	16,430,693	10,922,409

Table 4. Annual usage, operating cost, and protein unit price of alkaline extraction (Ctrl), enzymatic extraction (Ctrl-Alc), and enzymatic reactive extrusion (Ex-Alc) processing plants.

Parameters	Alkaline Extraction		Enzymatic Extraction		Reactive Extrusion Extraction	
	Annual Use	Annual Cost ($)	Annual Use	Annual Cost ($)	Annual Use	Annual Cost ($)
Raw materials						
Canola meal (kg)	50,000,000	15,000,000	50,000,000	15,000,000	50,000,000	15,000,000
Alcalase (kg)	-	-	50,000	1,150,000	50,000	1,150,000
HCl (kg)	180,576	39,727	180,576	39,727	129,888	28,575
NaOH (kg)	220,000	99,000	220,000	99,000	182,000	81,900
Water (ton)	445,885	312,120	445,885	312,120	195,385	136,770
Sub-total raw materials		15,450,846 (43.6%) [a]		16,600,846 (37.9%)		16,397,245 (55.6%)
Waste discharge						
Solid waste disposal (ton)	103,942,000	6,236,520	101,257,000	6,075,420	92,528,900	5,551,734
Sub-total waste discharge		6,236,520 (17.6%)		6,075,420 (13.9%)		5,551,734 (18.8%)
Utilities						
Electricity (kWh)	9,844,677	639,904	9,908,078	644,025	31,997,512	2,079,838
Steam (MT)	695,229	11,818,893	756,463	12,859,871	253,832	4,315,144
Cooling water (MT)	-	-	6,176,391	6,176,391	-	-
Sub-total utilities		12,458,797 (35.2%)		19,680,287 (44.9%)		6,394,982 (21.7%)
Labor						
Total salaries		300,000		300,000		300,000
Labor burden		270,000		270,000		270,000
Sub-total labor		570,000 (1.6%)		570,000 (1.3%)		570,000 (1.9%)
Other overhead						
Maintenance		77,962		98,736		52.905
Property insurance		78,222		99,065		53.081
Sub-total other overhead		156,185 (0.4%)		197,802 (0.5%)		105,985 (0.4%)
Depreciation		558,731 (1.6%)		707,610 (1.6%)		470,388 (1.6%)
Total operating cost		39,623,239 (100%)		44,452,575 (100%)		33,182,341 (100%)
Unit cost of protein		$10.6/kg		$7.0/kg		$3.6/kg

[a] Number in parenthesis represents the total operating cost share of each category.

https://doi.org/10.3390/foods15030498

4. Conclusions

This study established a semi-continuous strategy for producing canola protein ingredients by integrating limited proteolysis during extrusion (*eREX*) with short-duration alkaline extraction. A key finding is that a 5 min extraction following *eREX* pre-treatment achieved protein recoveries comparable to those obtained with Alcalase-assisted extraction and greater than those from alkaline extraction alone, thereby reducing the extraction time from hours to minutes. Proteolysis and structural reorganization induced during *eREX* improved solubility, foaming stability, oil-binding capacity, and amino acid composition while maintaining high digestibility, although gelling capacity was reduced. Due to the short residence time and high solids content used during *eREX*, the process also substantially lowers water and energy consumption and decreases operating costs relative to conventional alkaline extraction. The estimated production cost and recommended selling price support the feasibility of scaling the process for industrial application. Because isoelectric precipitation was intentionally omitted, the resulting extract exhibited lower purity than a protein isolate. However, it is suitable as a protein concentrate that also delivers a moderate level of dietary fiber. Further optimization of extrusion conditions and enzyme dosage may enhance protein purity and recovery. A comprehensive assessment of alternative protease systems is recommended to evaluate their influence on process efficiency, composition, and functional performance of the extracted proteins from canola meal and other oilseed materials.

Supplementary Materials: The following supporting information can be downloaded at: https://www.mdpi.com/article/10.3390/foods15030498/s1, Table S1. Operating costs applied in the study. Figure S1. SDS-PAGE profile of canola meal proteins extracted with varying Alcalase concentration. Figure S2. Enlarged photographs of canola meal extrudates at 55% moisture content without and with Alcalase. Figure S3. SEM micrographs of canola meal extracts of untreated control (Ctrl) and enzymatic reactive extrudates (Ex-Alc) at 65% moisture. Figure S4. Soluble protein content of non-extruded controls (Ctrl) and extrudates (Extr) at 5, 60, and 120 min of alkaline extraction at pH 9, both without and with Alcalase. Figure S5. Photographs of emulsions formed with canola meal extracts at (A) day 0 (B) day 1. Figure S6. Photographs of emulsions formed with canola meal extracts at (A) day 3 (B) day 7. Reference [60] is cited in the Supplementary Materials.

Author Contributions: S.G.: Investigation, Data curation, Methodology, Formal analysis, Writing—original draft. E.C.G.H.: Investigation, Formal analysis, Methodology. X.S.: Formal analysis, Review & editing: J.C.: Investigation, Methodology. F.S.M.-G.: Formal analysis, Methodology, Writing—review. Q.J.: Formal analysis, Methodology, Writing-original draft. D.C.: Formal analysis, Methodology, Writing—original draft. Conceptualization, Visualization, Supervision, Project administration, Methodology, Writing—review & editing, Funding acquisition. All authors have read and agreed to the published version of the manuscript.

Funding: This project has been funded by The Good Food Institute, grant number 2024-67017-42813.

Institutional Review Board Statement: Not applicable.

Informed Consent Statement: Not applicable.

Data Availability Statement: The original contributions presented in this study are included in the article/Supplementary Material. Further inquiries can be directed to the corresponding author.

Acknowledgments: I would like to thank Kwan You, Paul Alba, Lai Wei, and Enning Zhou for their help.

Conflicts of Interest: The authors declare no conflicts of interest.

https://doi.org/10.3390/foods15030498

Abbreviations

The following abbreviations are used in this manuscript:

Alc	Alcalase
Ctrl	Control
Ctrl-Alc	Protein extract from Alcalase-assisted extraction
DH	Degree of hydrolysis
EC	Emulsion capacity
ES	Emulsion stability
Ex-Ctrl	Protein extract from extruded canola meal
Ex-Alc	Protein extract from extruded canola meal containing Alcalase
FC	Foaming capacity
FS	Foaming stability
H_0	Surface hydrophobicity
pI	Isoelectric point
SDS-PAGE	Sodium dodecyl sulfate–polyacrylamide gel electrophoresis
TEA	Technoeconomic analysis
TCI	Total capital investment
WHC	Water-holding capacity
OBC	Oil-binding capacity

References

1. Good Food Institute. Reducing the Price of Alternative Proteins. Available online: https://gfi.org/reducing-the-price-of-alternative-proteins/ (accessed on 21 October 2025).
2. Jarunglumlert, T.; Chantanuson, R.; Hayashi, R.; Katano, Y.; Kusakari, T.; Nagamine, S.; Matsumiya, K.; Kobayashi, T.; Nakagawa, K. Techno-Economic Assessment of Plant-Based Meat Analogue Produced by the Freeze Alignment Technique. *Future Foods* **2023**, *8*, 100269. [CrossRef]
3. Hadidi, M.; Aghababaei, F.; Gonzalez-Serrano, D.J.; Goksen, G.; Trif, M.; McClements, D.J.; Moreno, A. Plant-Based Proteins from Agro-Industrial Waste and by-Products: Towards a More Circular Economy. *Int. J. Biol. Macromol.* **2024**, *261*, 129576. [CrossRef]
4. United States Department of Agriculture Oilseeds: World Markets and Trade. Available online: https://www.fas.usda.gov/data/production/commodity/2226000 (accessed on 21 October 2025).
5. Tan, S.H.; Mailer, R.J.; Blanchard, C.L.; Agboola, S.O. Canola Proteins for Human Consumption: Extraction, Profile, and Functional Properties. *J. Food Sci.* **2011**, *76*, R16–R28. [CrossRef]
6. Li, K.; Peng, D.; Shao, J.; Huang, F.; Jin, W.; Wan, X.; Yang, J.; Deng, Q. The Digestibility of Rapeseed Protein Isolate Prepared by Salt and Alkali Extraction: The Importance of Protein Composition. *Food Chem.* **2025**, *493*, 145852. [CrossRef]
7. Tang, Y.R.; Ghosh, S. A Review of the Utilization of Canola Protein as an Emulsifier in the Development of Food Emulsions. *Molecules* **2023**, *28*, 8086. [CrossRef]
8. Wanasundara, J.P.D.; McIntosh, T.C.; Perera, S.P.; Withana-Gamage, T.S.; Mitra, P. Canola/Rapeseed Protein-Functionality and Nutrition. *OCL* **2016**, *23*, D407. [CrossRef]
9. Ghodsvali, A.; Khodaparast, M.H.H.; Vosoughi, M.; Diosady, L.L. Preparation of Canola Protein Materials Using Membrane Technology and Evaluation of Meals Functional Properties. *Food Res. Int.* **2005**, *38*, 223–231. [CrossRef]
10. Kumar, M.; Tomar, M.; Potkule, J.; Verma, R.; Punia, S.; Mahapatra, A.; Belwal, T.; Dahuja, A.; Joshi, S.; Berwal, M.K.; et al. Advances in the Plant Protein Extraction: Mechanism and Recommendations. *Food Hydrocoll.* **2021**, *115*, 106595. [CrossRef]
11. Sari, Y.W.; Mulder, W.J.; Sanders, J.P.M.; Bruins, M.E. Towards Plant Protein Refinery: Review on Protein Extraction Using Alkali and Potential Enzymatic Assistance. *Biotechnol. J.* **2015**, *10*, 1138–1157. [CrossRef]
12. Cavailles, J.; Vaca-Medina, G.; Wu-Tiu-Yen, J.; Labonne, L.; Evon, P.; Peydecastaing, J.; Pontalier, P.Y. Aqueous pretreatment of lignocellulosic biomass for binderless material production: Influence of twin-screw extrusion configuration and liquid-to-solid ratio. *Molecules* **2024**, *29*, 3020. [CrossRef]
13. Zeitoun, R.; Pontalier, P.-Y.; Maréchal, P.; Rigal, L. Twin-screw extrusion for hemicellulose recovery: Influence on extract purity and purification performance. *Bioresour. Technol.* **2010**, *101*, 9348–9354. [CrossRef]
14. Evon, P.; Vandenbossche, V.; Candy, L.; Pontalier, P.Y.; Rouilly, A. Twin-screw extrusion: A key tech-nology for the biorefinery. In *Biomass Extrusion and Reaction Technologies: Principles to Practices and Future Potential*; American Chemical Society: Washington, DC, USA, 2018; pp. 25–44.

https://doi.org/10.3390/foods15030498

15. Colas, D.; Doumeng, C.; Pontalier, P.Y.; Rigal, L. Twin-screw extrusion technology, an original solution for the extraction of proteins from alfalfa (*Medicago sativa*). *Food Bioprod. Process.* **2013**, *91*, 175–182. [CrossRef]
16. Colas, D.; Doumeng, C.; Pontalier, P.Y.; Rigal, L. Green crop fractionation by twin-screw extrusion: Influence of the screw profile on alfalfa (*Medicago sativa*) dehydration and protein extraction. *Chem. Eng. Process.* **2013**, *72*, 1–9. [CrossRef]
17. Chen, D.; Campanella, O.H. Limited enzymatic hydrolysis induced pea protein gelation at low protein concentration with less heat requirement. *Food Hydrocoll.* **2022**, *128*, 107547. [CrossRef]
18. Lamsal, B.P.; Murphy, P.A.; Johnson, L.A. Flaking and extrusion as mechanical treatments for enzyme-assisted aqueous extraction of oil from soybeans. *J. Am. Oil Chem. Soc.* **2006**, *83*, 973–979. [CrossRef]
19. Duque, A.; Doménech, P.; Álvarez, C.; Ballesteros, M.; Manzanares, P. Study of the bioprocess condi-tions to produce bioethanol from barley straw pretreated by combined soda and enzyme-catalyzed extrusion. *Renew. Energy* **2020**, *158*, 263–270. [CrossRef]
20. Xu, E.; Campanella, O.H.; Ye, X.; Jin, Z.; Liu, D.; BeMiller, J.N. Advances in conversion of natural bi-opolymers: A reactive extrusion (REX)–enzyme-combined strategy for starch/protein-based food processing. *Trends Food Sci. Technol.* **2020**, *99*, 167–180. [CrossRef]
21. Laplante, T.; Pham, S.; Patel, N.; Konuklar, G. Method of Manufacturing a Nutritional Powder with In Situ Protein Hydrolysis. WO2018125920A1, 27 December 2017.
22. Mullner, S.; Otto, R.; Weiss, A.; Regina, S.; Merck, K.; Graaf, L.A. Extruded Protein Hydrolysate, Method for Production and Use Thereof. WO2002036801A2, 10 May 2002.
23. Anh, T.L.Q.; Hoa, N.T.Q.; Nguyen, P.D.T.; Thanh, H.V.; Nguyen, P.B.; Anh, L.T.H.; Dao, D.T.A. Soybean Protein Extraction by Alcalase and Flavourzyme, Combining Thermal Pretreatment for Enteral Feeding Product. *Catalysts* **2020**, *10*, 829. [CrossRef]
24. Nielsen, S. *Food Analysis Laboratory Manual*, 4th ed.; Food Science Text Series; Springer: Cham, Switzerland, 2024; ISBN 978-3-031-44970-3.
25. Sluiter, A.; Bonnie, H.; Raymond, R.; Christopher, S.; Justin, S.; David, T.; Crocker, D.L.A.P. *Determination of Structural Carbohydrates and Lignin in Biomass*; National Renewable Energy Laboratory: Golden, CO, USA, 2008.
26. Van Handel, E. Rapid Determination of Total Lipids in Mosquitoes. *J. Am. Mosq. Control. Assoc.* **1985**, *1*, 302–304.
27. Agboola, S.O.; Mofolasayo, O.A.; Watts, B.M.; Aluko, R.E. Functional Properties of Yellow Field Pea (*Pisum sativum* L.) Seed Flours and the in Vitro Bioactive Properties of Their Polyphenols. *Food Res. Int.* **2010**, *43*, 582–588. [CrossRef]
28. Wei, L.; Plazaola, V.G.M.; Limit, D.; Siddiqui, U.; Ilavsky, J.; Chen, D. Modulating the Gelling Behavior of Pea Proteins through Glutaminase-Induced Deamidation. *Food Hydrocoll.* **2026**, *171*, 111825. [CrossRef]
29. Siddiqui, U.; You, K.-M.; Wei, L.; Limit, D.; Martin-Gonzalez, F.S.; Simsek, S.; Chen, D. Influence of Glutaminase-Assisted Extraction at Varied pH on the Physicochemical Properties and Functionalities of Pea Proteins. *Appl. Food Res.* **2025**, *5*, 101059. [CrossRef]
30. Stone, A.K.; Karalash, A.; Tyler, R.T.; Warkentin, T.D.; Nickerson, M.T. Functional Attributes of Pea Protein Isolates Prepared Using Different Extraction Methods and Cultivars. *Food Res. Int.* **2015**, *76*, 31–38. [CrossRef]
31. Brodkorb, A.; Egger, L.; Alminger, M.; Alvito, P.; Assunção, R.; Ballance, S.; Bohn, T.; Bourlieu-Lacanal, C.; Boutrou, R.; Carrière, F.; et al. INFOGEST Static in Vitro Simulation of Gastrointestinal Food Digestion. *Nat. Protoc.* **2019**, *14*, 991–1014. [CrossRef]
32. Paes, M.C.D.; Maga, J. Effect of Extrusion on Essential Amino Acids Profile and Color of Whole-Grain Flours of Quality Protein Maize (Qpm) and Normal Maize Cultivars. *Rev. Bras. Milho E Sorgo* **2004**, *3*, 10–20. [CrossRef]
33. Bhaskar, B.; Ananthanarayan, L.; Jamdar, S.N. Effect of Enzymatic Hydrolysis on the Functional, Antioxidant, and Angiotensin I-Converting Enzyme (ACE) Inhibitory Properties of Whole Horse Gram Flour. *Food Sci. Biotechnol.* **2019**, *28*, 43–52. [CrossRef]
34. Gulati, P.; Rose, D.J. Effect of Extrusion on Folic Acid Concentration and Mineral Element Dialyzability in Great Northern Beans (*Phaseolus vulgaris* L.). *Food Chem.* **2018**, *269*, 118–124. [CrossRef]
35. Perera, S.; McIntosh, T.; Wanasundara, J. Structural Properties of Cruciferin and Napin of *Brassica napus* (Canola) Show Distinct Responses to Changes in pH and Temperature. *Plants* **2016**, *5*, 36. [CrossRef]
36. Di Lena, G.; Sanchez Del Pulgar, J.; Lucarini, M.; Durazzo, A.; Ondrejíčková, P.; Oancea, F.; Frincu, R.-M.; Aguzzi, A.; Ferrari Nicoli, S.; Casini, I.; et al. Valorization Potentials of Rapeseed Meal in a Biorefinery Perspective: Focus on Nutritional and Bioactive Components. *Molecules* **2021**, *26*, 6787. [CrossRef]
37. Rutherfurd, S.M.; Gilani, G.S. Amino Acid Analysis. *Curr. Protoc. Protein Sci.* **2009**, *58*, 11.9.1–11.9.37. [CrossRef]
38. Hurrell, R.F.; Carpenter, K.J. The estimation of available lysine in foodstuffs after Maillard reactions. *Prog. Food Nutr. Sci.* **1981**, *5*, 159–176.
39. Mookerjee, A.; Tanaka, T. Influence of enzymatic treatments on legume proteins for improved functional and nutritional properties: Expansion of legume protein utilization as food ingredients. *Curr. Opin. Food Sci.* **2023**, *49*, 100974. [CrossRef]
40. Osen, R.; Toelstede, S.; Eisner, P.; Schweiggert-Weisz, U. Effect of high moisture extrusion cooking on pro-tein–protein interactions of pea (*Pisum sativum* L.) protein isolates. *Int. J. Food Sci. Technol.* **2015**, *50*, 1390–1396. [CrossRef]
41. Ma, W.; Xie, F.; Zhang, S.; Wang, H.; Hu, M.; Sun, Y.; Zhong, M.; Zhu, J.; Qi, B.; Li, Y. Characterizing the Structural and Functional Properties of Soybean Protein Extracted from Full-Fat Soybean Flakes after Low-Temperature Dry Extrusion. *Molecules* **2018**, *23*, 3265. [CrossRef]

 https://doi.org/10.3390/foods15030498

42. Tacias-Pascacio, V.G.; Morellon-Sterling, R.; Siar, E.-H.; Tavano, O.; Berenguer-Murcia, Á.; Fernandez-Lafuente, R. Use of Alcalase in the Production of Bioactive Peptides: A Review. *Int. J. Biol. Macromol.* **2020**, *165*, 2143–2196. [CrossRef]
43. Gerzhova, A.; Mondor, M.; Benali, M.; Aider, M. Study of Total Dry Matter and Protein Extraction from Canola Meal as Affected by the pH, Salt Addition and Use of Zeta-Potential/Turbidimetry Analysis to Optimize the Extraction Conditions. *Food Chem.* **2016**, *201*, 243–252. [CrossRef]
44. Zhang, Y.; Romero, H.M. Exploring the Structure-Function Relationship of Great Northern and Navy Bean (*Phaseolus vulgaris* L.) Protein Hydrolysates: A Study on the Effect of Enzymatic Hydrolysis. *Int. J. Biol. Macromol.* **2020**, *162*, 1516–1525. [CrossRef]
45. Vioque, J.; Sánchez-Vioque, R.; Clemente, A.; Pedroche, J.; Millán, F. Partially Hydrolyzed Rapeseed Protein Isolates with Improved Functional Properties. *J. Am. Oil Chem. Soc.* **2000**, *77*, 447–450. [CrossRef]
46. Yust, M.D.M.; Pedroche, J.; Millán-Linares, M.D.C.; Alcaide-Hidalgo, J.M.; Millán, F. Improvement of Functional Properties of Chickpea Proteins by Hydrolysis with Immobilised Alcalase. *Food Chem.* **2010**, *122*, 1212–1217. [CrossRef]
47. Al-Ruwaih, N.; Ahmed, J.; Mulla, M.F.; Arfat, Y.A. High-Pressure Assisted Enzymatic Proteolysis of Kidney Beans Protein Isolates and Characterization of Hydrolysates by Functional, Structural, Rheological and Antioxidant Properties. *LWT* **2019**, *100*, 231–236. [CrossRef]
48. Ntone, E.; Van Wesel, T.; Sagis, L.M.C.; Meinders, M.; Bitter, J.H.; Nikiforidis, C.V. Adsorption of Rapeseed Proteins at Oil/Water Interfaces. Janus-like Napins Dominate the Interface. *J. Colloid Interface Sci.* **2021**, *583*, 459–469. [CrossRef]
49. García-Moreno, P.J.; Gregersen, S.; Nedamani, E.R.; Olsen, T.H.; Marcatili, P.; Overgaard, M.T.; Jacobsen, C. Identification of emulsifier potato peptides by bioinformatics: Application to omega-3 delivery emulsions and release from potato industry side streams. *Sci. Rep.* **2020**, *10*, 690. [CrossRef]
50. Guo, F.; Xiong, Y.L.; Qin, F.; Jian, H.; Huang, X.; Chen, J. Surface properties of heat-induced soluble soy protein aggregates of different molecular masses. *J. Food Sci.* **2015**, *80*, C279–C287. [CrossRef]
51. Delahaije, R.J.B.M.; Gruppen, H.; Giuseppin, M.L.F.; Wierenga, P.A. Towards Predicting the Stability of Protein-Stabilized Emulsions. *Adv. Colloid Interface Sci.* **2015**, *219*, 1–9. [CrossRef]
52. Ma, K.; Zhang, L.; Sun, X.; Chen, F.; Zhu, T. Correlationship between Self-Assembly Behavior and Emulsion Stabilization of Pea Protein-High Methoxyl Pectin Complexes Treated with Ultrasound at pH 2.0. *Ultrason. Sonochem.* **2023**, *100*, 106596. [CrossRef]
53. Cháirez-Jiménez, C.; Castro-López, C.; Serna-Saldívar, S.; Chuck-Hernández, C. Partial Characterization of Canola (*Brassica napus* L.) Protein Isolates as Affected by Extraction and Purification Methods. *Heliyon* **2023**, *9*, e21938. [CrossRef]
54. Von Der Haar, D.; Müller, K.; Bader-Mittermaier, S.; Eisner, P. Rapeseed Proteins—Production Methods and Possible Application Ranges. *OCL* **2014**, *21*, D104. [CrossRef]
55. Singh, R.; Koksel, F. Effects of Particle Size Distribution and Processing Conditions on the Techno-Functional Properties of Extruded Soybean Meal. *LWT* **2021**, *152*, 112321. [CrossRef]
56. Leonard, W.; Zhang, P.; Ying, D.; Nie, S.; Tindal, E.; Fang, Z. Transformation of Hempseed (*Cannabis sativa* L.) Oil Cake Proteome, Structure and Functionality after Extrusion. *Food Chem.* **2022**, *384*, 132499. [CrossRef]
57. Chen, D.; Zhu, X.; Ilavsky, J.; Whitmer, T.; Hatzakis, E.; Jones, O.G.; Campanella, O.H. Polyphenols weaken pea protein gel by formation of large aggregates with diminished noncovalent interactions. *Biomacromolecules* **2021**, *22*, 1001–1014. [CrossRef]
58. Ivanova, P.; Chalova, V.; Uzunova, G.; Koleva, L.; Manolov, I. Biochemical Characterization of Industrially Produced Rapeseed Meal as a Protein Source in Food Industry. *Agric. Agric. Sci. Procedia* **2016**, *10*, 55–62. [CrossRef]
59. Goertzen, A.D.; House, J.D.; Nickerson, M.T.; Tanaka, T. The Impact of Enzymatic Hydrolysis Using Three Enzymes on the Nutritional Properties of a Chickpea Protein Isolate. *Cereal Chem.* **2021**, *98*, 275–284. [CrossRef]
60. Humbird, D.; Davis, R.; Tao, L.; Kinchin, C.; Hsu, D.; Aden, A.; Schoen, P.; Lukas, J.; Olthof, B.; Worley, M.; et al. *Process Design and Economics for Biochemical Conversion of Lignocellulosic Biomass to Ethanol: Dilute-Acid Pretreatment and Enzymatic Hydrolysis of Corn Stover*; NREL/TP-5100-47764, 1013269; NREL: Golden, CO, USA, 2011. [CrossRef]

 https://doi.org/10.3390/foods15030498

MDPI AG
Grosspeteranlage 5
4052 Basel
Switzerland
Tel.: +41 61 683 77 34

Foods Editorial Office
E-mail: foods@mdpi.com
www.mdpi.com/journal/foods

www.ingramcontent.com/pod-product-compliance
Lightning Source LLC
Chambersburg PA
CBHW061624210926
51850CB00009B/2433